United States Marine Corps
Command and Staff College
Marine Corps University
2076 South Street
Marine Corps Combat Development Command
Quantico, Virginia 22134-5068

MASTER OF MILITARY STUDIES

Falling From Grace:
The German Airborne in World War II

Submitted in partial fulfillment
of the requirements for the degree of
Master of Military Studies

Author: Chris Mason

AY 2000-2001

Mentor: Dr. Charles D. McKenna, PhD
Approved: _____
Date: _____

Mentor: Colonel James McCarl, USA
Approved: _____
Date: _____

Report Documentation Page

Report Date 2002	Report Type N/A	Dates Covered (from... to) -

Title and Subtitle Falling From Grace: The German Airborne in World War 2	Contract Number
	Grant Number
	Program Element Number
Author(s)	Project Number
	Task Number
	Work Unit Number
Performing Organization Name(s) and Address(es) Joint Military Operations Department Naval War College 686 Cushing Road Newport, RI 02841-1207	Performing Organization Report Number
Sponsoring/Monitoring Agency Name(s) and Address(es)	Sponsor/Monitor's Acronym(s)
	Sponsor/Monitor's Report Number(s)

Distribution/Availability Statement
Approved for public release, distribution unlimited

Supplementary Notes

Abstract
Nazi Germany pioneered the use of airborne forces in the 20th century and used them effectively early in the war as part of integrated, combined-arms offensives. Yet the German airborne branch literally self-destructed in 1941. What happened, how did the Germans react, and what historical insights in the use of airborne forces can modern day planners derive from the German experience?

Subject Terms

Report Classification unclassified	Classification of this page unclassified
Classification of Abstract unclassified	Limitation of Abstract UU

Number of Pages
58

Table of Contents

Part I:	Introduction	1
Part II:	Genesis: 1936-1938	1
Part III:	The Use of Paratroops: The First Great Debate, 1936-1938	5
Part IV:	Deployment Doctrine: The Second Great Debate, 1939	9
Part V:	Testing the Doctrine: Early Airborne Operations, 1939-1940	14
Part VI:	1941: The Turning Point: Crete or Malta?	17
Part VII:	"Pride Goeth Before Destruction": Crete 1941	19
Part VIII:	The Battle Plan	23
Part IX:	Pyrrhic Victory	25
Part X:	The Aftermath of Crete: Two Year Hiatus, 1941-1943	30
Part XI:	A Fighting Reputation	35
Part XII:	Final Airborne Operations: 1943-1945	38
Part XIII:	What Might Have Been	42
Part XIV:	Final Thoughts	44

Disclaimer

The opinions and conclusions expressed herein are those of the individual student author and do not necessarily represent the views of either the Marine Corps Command and Staff College or any other governmental agency. References to this study should include the foregoing statement.

Executive Summary

Title: Falling from Grace: German Airborne Forces in World War II

Author: M. Chris Mason, FS-3, United States Department of State

Thesis: Nazi Germany pioneered the use of airborne forces in the 20th century and used them effectively early in the war as part of integrated, combined-arms offensives. Yet the German airborne branch literally self-destructed in 1941. What happened, how did the Germans react, and what historical insights in the use of airborne forces can modern day planners derive from the German experience?

Discussion: In the late 1930's, an aggressive and innovative rearmament program in Nazi Germany gave rise to the tactics of vertical envelopment. Pioneering the use of gliders as troop carriers, parachutists, and the air landing of reinforcements to exploit tactical success, the German *Wehrmacht* used the new technique of airborne warfare with startling success as part of the *Blitzkrieg* campaign against the Low Countries and France in 1940.

When the tactical doctrine used to seize bridges, strong points and road junctions in *Fall Gelb* was transferred to the seizure of an entire island that was heavily defended in 1941, however, the German airborne effectively committed suicide. In ten days in May 1941, half the airborne forces in the entire German army were killed or wounded on Crete. Hitler wrongly ascribed the disaster to a playing out of the surprise factor, and banned further parachute operations until 1943.

The right conclusions were arrived at by the commander of the German airborne himself, General Kurt Student, in post-battle analysis. His own insistence on faulty tactics was devastating. Although they remained a potent and professional force, Hitler's effective ban on the future use of airborne forces lasted until 1943, when it was clear the Allies still very much considered paratroops a viable form of warfare. By then, Germany's ability to conduct airborne operations on a significant scale had long since passed.

Conclusion: The German innovation of vertical envelopment in the 1930's was as revolutionary to modern military tactics as the simultaneous development of the integrated combined arms offensive known today as the *Blitzkrieg*. In putting Billy Mitchell's ideas into practice, *Luftwaffe* General Student demonstrated vision, innovative thinking and practical military skill. Poor intelligence and reliance on his "spreading oil drops" tactics for the deployment of his paratroopers, the *Fallschirmtruppe,* on Crete, however, led directly to their removal as a significant weapon from the German arsenal in World War II. Nevertheless, Student proved that airborne troops have unique capabilities as a force multiplier in both offensive and defensive warfare. The German experience, which also demonstrated the limitations of airborne forces, was studied in depth by the U.S. Army after the war and incorporated into airborne doctrine.

"Where is the prince who can afford to so cover his country with troops for its defense, as that 10,000 men descending from the clouds might not, in many places, do an infinite deal of mischief before a force could be brought to repel them?"

Benjamin Franklin, 1784[1]

Part I: Introduction

In the late 1930's, an aggressive and innovative rearmament program in Nazi Germany gave rise to the tactics of vertical envelopment. *Luftwaffe* General Kurt Student pioneered the use of gliders and airborne forces, and the *Wehrmacht* used them effectively early in the war as part of integrated, combined-arms offensives. Yet the German airborne branch literally self-destructed in May 1941 during the invasion of the island of Crete. This paper examines the development of German airborne doctrine and addresses the questions of what happened on Crete, how the Germans reacted, and what historical insights modern day planners can derive from the German experience.

Part II: Genesis 1936-1938

The idea was risky, but if it worked, it could make it possible to overrun all of France in a matter of weeks, a problem that had been puzzling military planners for years. About 600 aircraft were needed for the operation, in which a division of light infantry would be dropped by parachute behind enemy lines. By seizing key terrain objectives, disrupting enemy communications, and sowing panic in the enemy's rear, they would pave the way for a coordinated offensive through the sector. The operation was approved and scheduled for execution in 90 days. The date was October 17, 1918, the Armistice was a month away, and the planner of "Operation Panic Party"

---

[1] Ryan, Cornelius. *A Bridge Too Far*. Touchstone Books. New York. 1974. p. 122.

was the visionary American air power pioneer, Major General William "Billy" Mitchell, at that time the commander of American Air Forces in Europe.[2]

Even then, the idea of vertical envelopment[3] was an old one. Thirty years before, in 1889, the man who would become the Chief of the German General Staff, General Graf von Schlieffen, observed a parachute demonstration by American balloonist Charles Leroux at Schöneberg near Berlin. In his report, von Schlieffen wrote: "if it were only possible to steer these things, parachutes could provide a new means of exploiting surprise in war."[4]

The only verified use of parachutes in World War I, however, was to save pilots and balloonists' lives.[5] After the Armistice in November 1918, centuries of martial dreams of leaping over enemy positions in war again lay nearly dormant. Apart from some testing of parachutes for pilots in the 1920's, the United States dismissed military parachuting.[6] Mitchell, a human bulldog with an idea, however, had not given up entirely. When he first made public his plan for Operation Panic Party in a speech in New York in 1919,[7] he was considered by many people to have lost touch with reality,[8] but in the 1920's he organized a demonstration jump by six U.S. Army parachutists at Kelly Field in Texas. General Hap Arnold timed them as they set up and fired a machine gun within three minutes after landing.[9] In the climate of postwar disarmament the idea went nowhere.

---

[2] Mitchell, William. *Memoirs of World War I*. Random House. New York. 1960. p. 268.
[3] The earliest use of this term found was in an article entitled "Air Infantry Training" in the July-August 1940 issue of the U.S. Army magazine *Infantry Journal* (p. 329). The article, which describes the German airborne training of the period in detail, also envisioned the future development of "a large helicopter plane..that would carry…fifty men and their equipment."
[4] Kuhn, Volkmar. *German Paratroops in World War II*. Ian Allen Ltd, London, 1978, p. 7.
[5] Only German aircraft pilots used parachutes, and then only late in the war. Balloon observers on both sides were equipped with automatic opening parachutes, a wise precaution as the balloons were filled with hydrogen.
[6] Devlin, Gerard. *Paratrooper!* St. Martin's Press, NY 1979. p. 32.
[7] Devine, Isaac. *Mitchell, Pioneer of Air Power*. Duell, Sloan and Pearce, NY. 1943. p.151.
[8] Kurowski, Franz. *Deutsche Fallschirmjäger 1939-1945*. Edition Aktuell, Frankfurt, c. 1990, p. 12.

Only in the Soviet Union was any serious consideration given to Mitchell's concept in the post-war years. In 1930, the year Mitchell's book *Skyways* was published, the Red Army deployed a small section of paratroops in maneuvers for the first time in history. Again in 1931, a section of seven men participated in Soviet summer maneuvers.[10] Later that year, Soviet Air Force commander Marshall Schtscherbakow toured the Maginot Line with French Marshall Petain, who boasted, "these fortresses will prevent any incursion into our territory." The Soviet Marshall is said to have replied to a shaken Petain, "if the enemy pursues General Mitchell's ideas, and parachutes over them…fortresses like these may well be superfluous in the future."[11] The French military then dabbled briefly with the concept, creating two companies of paratroopers in 1936, but the experiment was dismissed by the French General Staff as "a circus act" and abolished before the war started.[12]

Hitler's rise to power in 1933 initiated a shift in European military power. He moved to circumvent the restrictions of the Versailles Treaty in a number of ways. For example, although Germany was prohibited from having an air force, Hitler instituted a large civil sport gliding organization that provided thousands of future pilots with the fundamentals of flight training.[13] Meanwhile, German pilots secretly trained in powered flight in the Soviet Union.[14] The 100,000-man limit on the army was also sidestepped. For example, many police formations were paramilitary in nature, and received extensive military training, including the Prussian

---

[9] Arnold, General H.H. *Winged Warfare.* Harper & Brothers. New York. 1941. p. 56.

[10] Kuhn, p.8.
[11] Ibid.
[12] Piehl, Hauptmann. *Ganze Männer.* Verlagshaus Bong & Co. Leipzig. 1943. p. 41.
[13] The organization was called the *Deutscher Luftsport Verband*, or "German Air Sports Society." Bender, Roger, *Air Organizations of the Third Reich.* R. James Bender Publishing. Mountain View, California. 1967. p. 8.
[14] Cooper, Matthew. *The German Air Force 1933-1945, an Anatomy of Failure.* Jane's Publishing, Cooper and Lucas, Ltd. London. 1981. p. 2. The Chief of the "*Flugtechnik*" section responsible for clandestine pilot training and secret aircraft development was *Hauptmann* Kurt Student.

"*Polizei Regiment General Göring*,"[15] named for World War I flying ace Hermann Göring, now a senior Nazi leader.

1935 was a watershed. On April 1st, the Germans made their first move toward joining the Soviets in airborne development. A battalion of the Göring Regiment was detached from the ranks of the police and formally constituted as a military unit.[16] Meanwhile, Soviet development of airborne forces continued. Surprisingly, during the early 1930's the maneuvers of the Red Army under Joseph Stalin were open to western observers and military attachés. In the summer of 1935, for example, both German and British observers watched the annual Soviet training exercises in the Ukraine. This time, 1,000 paratroops were dropped as part of the exercise, followed by 5,000 more brought by airlift into the "captured" area. British Colonel Archibald Wavell, later the Commander in Chief of British forces in the Middle East, was present, and submitted a report to the British War Office, writing: "If I had not been an eye-witness to this event I would never have believed that such an operation was possible."[17]

But while the British slept on Wavell's observations, the Nazis did not. On October 1, 1935, a battalion of the Göring regiment was transferred to the new German Air Force, the *Luftwaffe*. The battalion commander, a Lieutenant Colonel Jacoby, was informed by now Air Minister Göring himself that he was to call for volunteers "to train as paratroops...the Hermann Göring parachute battalion will form the nucleus of a German paratroop corps."[18] The formal

---

[15] Otte, Alfred. *The HG Panzer Division*. Schiffer Publishing Ltd. West Chester, PA. 1989. p. 5.
[16] Busch, *Die Fallschirmjäger Chronik 1935-1945*. Podzun Pallas Verlag. Friedberg, 1983. p. 13.
[17] Ailsby, Christopher. *Hitler's Sky Warriors*. Brassey's, Inc. Dulles, VA. 2000. p. 18.
[18] Kuhn, p. 9.

establishing order, *L.A. 262/36 gIII 1A*, was published January 29, 1936.[19] Parachutes and specialist equipment were designed and requisitioned.[20] For training, a parachuting school was established at Stendal, Germany in early 1936,[21] and on May 4th, the first class of trainees reported for instruction. The German airborne, the *Fallschirmtruppe*, began to grow.

Part III: The Use of Paratroops: The First Great Debate, 1936-1938

The world they grew into was one of breakneck German military rearmament and Byzantine Nazi politics. Personal power struggles were an omnipresent element of the bizarre Nazi system of government as an array of henchmen competed with one another to expand their personal influence and establish their own mini-empires of power. All sought favor in a zero-sum game of prestige in Hitler's court, and at the center of it all was the German Air Force Commander in Chief, *Reichsmarshall* Hermann Göring. In the 1930's, Germany, like Great Britain, had opted for a separate air force, and Göring was its absolute master. Unlike Great Britain, however, Germany did not build an air fleet for strategic bombing, as advocated by Italian air power theorist Giulio Douhet, opting instead for an air force designed to support the army in ground operations.[22]

Like almost everything else in Germany, the early German airborne was affected by the machinations of Nazi power. As the visibility of this new weapon grew in the *Luftwaffe*, so did the inter-service rivalry around it. Not wanting to be left out of this new kind of warfare, the German army (*Heer*) created a "Parachute Infantry Company" (*Fallschirm Infanterie Kompanie*)

---

[19] Otte, p. 14.
[20] Edwards, Roger. *German Airborne Troops*. Doubleday & Co. Garden City, New York. 1974. pp 24-25.
[21] Busch, p. 276.
[22] Magenheimer, Heinz. *Hitler's War, Germany's Key Strategic Decisions 1940-1945*. Cassel & Co., London 1997. p. 22. For detailed discussion of German air doctrine, see also Griess, Thomas, series editor. *The Second World War: Europe and the*

of its own in April 1937. Other than utilizing air force training facilities for parachute instruction, this force had no connection to the air force paratroops. It, too, grew rapidly, to comprise over 900 trained paratroops by mid-1938. These men wore army uniforms, were supplied by the army supply system, and wore a different parachutist badge from the *Luftwaffe* design on their uniforms.[23] Even the *SS* experimented briefly with a platoon of parachutists in the *Germania* Regiment in 1937.[24]

The *Luftwaffe* and the *Heer* also had very different ideas about employing paratroops. The *Luftwaffe* High Command envisioned paratroops essentially as saboteurs - - small groups of highly trained commandos who would parachute behind enemy lines to destroy key targets in the enemy's lines of communication.[25] They were to be an extension of the aircraft as a means of destroying targets on the ground, and were correspondingly trained primarily as demolition engineers. The *Heer* took an entirely different view. Paratroops were seen as conventional infantry who simply got to battle a different way. Their training focused on infantry tactics with an emphasis on the methods of the assault groups (*Sturmtruppen*) of the First World War.[26] Throughout 1937 and early 1938, the two paratroop forces developed along these divergent lines. Paratroopers of both branches participated successfully in the annual fall maneuvers of October 1937 with battalion-sized units and with missions reflecting the two different tactical philosophies.[27]

Mediterranean. Department of History, United States Military Academy, West Point. Avery Publishing Group, New Jersey. 1989. pp. 68-69.

[23] Schlicht, Adolf and Angolia, John. *Uniforms and Traditions of the German Army 1939-1945, Volume 1*. R. James Bender Publishing. San Jose. 1984. pp. 281-284.
[24] Milius, Siegfried. *Fallschirmjäger der Waffen-SS im Bild*. Munin Verlag GmbH. Osnabrück. 1986. p. 16.
[25] Busch, Erich. p. 20.
[26] Peters, Klaus. *Fallschirmjäger Regiment 3*. R. James Bender Publishing. San Jose, CA. 1992. p. 101.

By mid-1938, the pompous and ambitious Göring was near the apogee of his power and influence within the Nazi hierarchy. Paratroopers were an element of power, and Göring wanted exclusive control of them. Because the *Luftwaffe* would control the transport assets necessary to deliver paratroops to battle, Göring made his case to Hitler that for maximum efficiency, the *Luftwaffe* should have sole command and control of all paratroops. Although this model was not followed by any other power in World War II, Göring was able to win the point because of his position in the Machiavellian politics of personality in the Third Reich. In July 1938, after a little more than a year of separate existence, the army paratroops were transferred *en masse* into the *Luftwaffe*, forming the *II. Bataillon* of an envisioned 1st Parachute Regiment.

From the beginning, the paratroops had attracted the adventurers and the freethinkers in the German officer corps. But their development was now well advanced from the entrepreneurial "start-up" phase of their birth. The issue of air force vs. army command and control had been resolved, the paratroops had proven their potential in exercises and it was, in a sense, time for adult supervision. To take the concept from a novel military experiment to an integrated, operational element of the *Wehrmacht,* however, it would now be necessary to overcome a great deal of skepticism at senior levels. For the task, the *Luftwaffe* turned to General Kurt Arthur Benno Student, who took command of the airborne forces on July 1, 1938.[28]

In some ways, Student was an unlikely choice. Born in 1890, he was a flying ace in WWI,[29] not an infantry officer. After the war, Student had had an otherwise undistinguished career in largely

---

[27] Von Hove, Alkmar. *Achtung Fallschirmjäger, Eine Idee Bricht Sich Bahn.* Druffel-Verlag. Freising. 1954. pp. 24-25.
[28] Kuhn, p. 16.
[29] Student had five aerial victories before being seriously wounded in 1918. MacDonald, Callum. *The Battle of Crete.* Macmillan. New York. 1993. p. 7.

administrative billets in the 100,000-man *Reichsheer* of the 1920's and 1930's. He had limited experience with ground forces, commanding only at the company and battalion levels, first as a captain, then as a major, in the 2nd and 3rd Infantry Regiments from 1928-33.[30] But he was the right man for the job. Student, too, had been present at the Soviet airborne maneuvers of 1935, and needed no convincing of the value of the paratroops. Personally brave to the point of recklessness, he was selfless and dedicated, had a flair for innovation, and proved to be a brilliant administrator, organizer and logistician. As a professional officer, Student refused throughout the war to join the Nazi party [31] despite the negative effect this refusal had on promotion. As soon as he took command, he moved quickly to end the tactical debate about the role of the *Fallschirmjäger*. Because of his instrumental role in the development of German doctrine, Student's thinking on this matter, written after the war, is worth quoting here at some length:

> I could not accept the saboteur force concept. It was a daredevil idea but I did not see minor operations of this kind as worthwhile. They wasted individual soldiers and were not tasks for a properly constituted force. The chances of getting back after such missions appeared to be strictly limited…Casualties are inevitable in war, but soldiers must be able to assume a real chance of survival, and an eventual return home…From the very beginning, my ideas went much, much further. In my view airborne troops could become a battle-winning factor of prime importance. Airborne forces made three-dimensional warfare possible in land operations. An adversary could never be sure of a stable front because paratroops could simply jump over it and attack from the rear when and where they decided. There was nothing new about attacking from behind of course, such tactics have been practiced since the beginning of time and proved both demoralizing and effective. But airborne troops provided a new means of exploitation and so their potential in such operations was of incalculable importance. The element of surprise was an added consideration; the more paratroops dropped, the greater the surprise.[32]

The first great debate about airborne doctrine was thus resolved. Paratroops would be employed as light infantry[33]. But enormous doctrinal questions still lay ahead.

---

[30] Farrar-Hockley, A.H. *Student*. Ballantine Books. New York. 1973. p. 34.
[31] Barnett, Correlli, ed. *Hitler's Generals*. William Morrow & Co. New York. 1989. p. 478.
[32] Kuhn, p. 16.

Part IV: Deployment Doctrine: The Second Great Debate, 1939

Student went to work with a will in 1938. He recruited Germany's best and brightest to join the new military branch.[34] Standards were high. Every man had to be a volunteer, and only one pre-war applicant in four completed the training.[35] Student's work enjoyed the personal support of Hitler himself, for whom the paratroops represented exactly the kind of warfare he favored. Hitler also knew the elite nature and special uniforms of the paratroopers would appeal to the German public. It was no accident that the *Fallschirmjäger* were selected to lead the annual *Wehrmacht* parade in Berlin in 1939.[36]

Student expanded the concept of the airborne division to include three different types of forces. The first would be the paratroops, but Student quickly grasped the fact that Germany's resources would not permit the training of large numbers of parachutists. Three additional parachute training schools were set up during the course of the war, but shortly before the last one closed down operations in October 1944,[37] General Student reported to Hitler that a total of only 30,000 parachutists had been trained in the eight years the centers were open.[38] In the first five years, from May 1936 to May 1941, when the airborne invasion of Crete was launched, just 8,000 men had completed the training, an average of 130 trained parachutists per month. (After Crete, with the establishment of additional schools and the shortening of the course of instruction from eight weeks to three, training output increased to an average of 580 men per month.)

---

[33] Ironically, the saboteur role was soon resurrected by the army in the form of the special operations Brandenburg Regiment, which contained a high percentage of Germans who had lived outside of Germany before the war and spoke a number of foreign languages. See Brockdorff, Werner. *Geheim Kommandos des Zweiten Weltkrieges*. Weltbild Verlag GmbH, Augsburg. 1993.

[34] Several Olympic champions and even world heavyweight boxing champion Max Schmelling volunteered.
[35] McDonald, p. 16.
[36] Hickey, Michael. *Out of the Sky*. Charles Scribner's Sons. New York. 1979. pp.21-22.
[37] Busch, p. 279.

The second type of forces to be employed was glider troops. Student and his Chief of Staff, *Oberst* (Colonel) Hans Jeschonnek,[39] were both glider enthusiasts, but as Jeschonnek told Student, "nobody seems much interested in them."[40] Student was, however. Gliders were initially developed in the Third Reich to train pilots. In the early 1930's, some thought had been given to using them as cargo transporters, but with the rapid rise of the *Luftwaffe* they had been languishing in a developmental limbo ever since. In 1938, Student had an inspiration: The glider might be used to bring men to the battlefield. The means were already at hand in the *Gotha DFS 230*. Developed in 1933 as a cargo glider,[41] the *DFS 230* glider flew well in *Luftwaffe* tests but had been relegated by bureaucratic indifference to a test section run by a couple of lieutenants at a remote *Luftwaffe* air station.

After a test ride at the backwater *Ausbildungskommando für Lastensegelflug* (Training Command for Cargo Gliding), Student was immediately convinced the glider had offensive potential.[42] The DFS230 could bring ten fully armed men down together in one place ready for immediate action,[43] and their arrival could be nearly silent if detached from their tow planes at a distance from their targets. Furthermore, as Student was well aware, Germany already possessed most of the best glider pilots in the world as a result of Hitler's aggressive promotion of gliding as a Nazi-sponsored national sport throughout the 1930's. He moved quickly to bring the glider program under his command and, against considerable initial bureaucratic resistance from the

---

[38] Department of the Army. *Historical Study No. 20-232. Airborne Operations, A German Appraisal.* October 1951. p. 13.

[39] Jeschonnek's military career rapidly outpaced Student's. By the time of the Crete operation in May 1941, Jeschonnek had been promoted to Chief of Staff of the *Luftwaffe*. (Beevor, Anthony. *Crete, the Battle and the Resistance.* West View Press. Boulder. 1994. p. 73.) He committed suicide in August 1944. (Farrar-Hockley p. 127.)

[40] Kuhn, p. 17.

[41] Ibid.

[42] Student and Götzel, *Generaloberst Kurt Student und Seine Fallschirmjäger.* Podzun Pallas Verlag, Friedberg, 1980. p. 41.

[43] Nowarra, Heinz. *German Gliders in World War II.* Schiffer Publishing. West Chester, PA. 1991. p. 47.

Luftwaffe establishment,[44] developed them into a sharp offensive weapon. Full credit for the innovation of the troop-carrying glider goes to Kurt Student, who had the intellect to recognize a whole new weapon of war, the organizational skill to develop it, and the rank to make it happen.

The third type of forces in the German airborne model would be "air landing" troops.[45] These would be conventional infantry with extra training in emplaning and deplaning from aircraft. From the modern perspective, this notion seems quaint, but it should be remembered that in the mid-1930's less than one German in a thousand had ever been inside an airplane (much less contemplated jumping out of one in flight). Student and the man who replaced Jeschonnek as his deputy, Heinz Trettner, believed some familiarization with the process would be necessary.[46] The army thus re-designated the 22nd Infantry Division the "22nd Air Landing Division," trained the troops in boarding transports and even took them on short familiarization flights.[47]

Student therefore saw an airborne operation as a three-stage process: First, a small force of glider-borne troops would land close to key tactical elements such as communications centers, anti-aircraft batteries, road junctions and bridges, and neutralize them. Minutes later, paratroops would be dropped to secure other key geographical and military features and reinforce the glider troops. Finally, if an airfield had been secured, the "air-landing" troops would be flown into the airhead to expand outwards and eventually link up with conventional land-based forces.[48]

---

[44] Farrar-Hockley, A.H. *Student*. Ballantine Books. New York. 1973. pp. 52-53.
[45] Lucas, James. *Storming Eagles*. Arms and Armor Press. London. 1988. p. 12.
[46] Kuhn, p.16-17. Trettner remained as Student's deputy for three years. He ended the war as Commanding General of the *4. Fallschirm Division* in Italy.
[47] Mitcham, Samuel W., Jr. *Hitler's Legions, The German Army Order of Battle*. Stein and Day. New York. 1985. pp. 58-59.
[48] Department of the Army *Historical Study No. 20-232*, p. 9.

Indeed, Student and his staff demonstrated a remarkable grasp of the limitations and advantages of paratroops at a very early stage in their development. They understood that airborne operations were a new and valuable part of a combined arms offensive, but because of supply problems, they would have a very short combat staying power of three to four days at most, even with aerial re-supply, and would need to be relieved quickly by a link-up with conventional armored infantry forces. Student also clearly appreciated that by their nature, paratroops and air-landed troops were very light infantry. Apart from a handful of experimental drops of 3.7 cm lightweight anti-tank guns[49], they would have no weapon heavier than a 10 cm mortar[50] until late in the war. The problems of fire support and antitank defense were never fully resolved by the German airborne during the war, despite experimentation with air-dropping of artillery pieces and anti-tank guns, and the air landing (but not dropping) of 7.5 cm mountain guns and tracked motorcycles as prime movers.[51]

One additional logistical problem addressed by German airborne planners was how to get individual equipment, including weapons, medical supplies and radio sets into action. The handling of this seemingly minor detail was to have disastrous consequences for the German paratroopers. Unable to solve the safety issues of parachutists jumping into battle with the long K98 rifle and other weapons to their satisfaction, equipment designers came up with a "Rube Goldberg" solution.[52] Rifles, submachine guns, and all the ammunition and heavier weapons would be packed into separate equipment containers and parachute dropped from the bomb bays

---

[49] The 3.7 cm *Panzer Abwehr Kannone*, or *PAK*s, were assigned to the 14th Company of each regiment. They were to be air-dropped under a cluster of five parachutes called a "*Fünfling*," along with motorcycles with sidecars as their prime movers.
[50] Each regiment had one company (the 13th) of heavy 10 cm mortars, designated the *Nebelwerfer 35*, which were probably originally designed for chemical warfare. In addition, the heavy weapons companies of each battalion (the 4th, 8th and 12th companies) were equipped with 81 mm mortars in addition to heavy machine guns. However, German tactical doctrine called for only 24 rounds to be dropped with each mortar tube. Von Roon, Arnold. *Die Bildchronik der Fallschirmtruppe 1935-1945*, p. 78 and Farrar-Hockley, p. 50. Sergeant Ernst Simon, *Fallschirmjäger Regiment 1*. Interview by the author, tape recording, Hamburg, Germany, 1995.
[51] Department of the Army *Historical Study No. 20-232*. p. 13.
[52] Veranow, Michael, ed. *The Third Reich at War*. Galahad Books. New York. 1999. p. 173.

of the same JU52 aircraft used to transport and drop the paratroops.[53] All that each individual paratrooper would have on his person upon landing would be his 9mm pistol with two eight round magazines,[54] and several concussion grenades.[55] The immediate need for the individual paratrooper after landing was not, therefore, to link up with his comrades and rapidly engage the enemy, but rather to find the weapons canister. It was to prove a terrible mistake.

A far greater mistake, with even more far-reaching consequences, was made by Student himself in the development of tactical doctrine. As the development of the force and its equipment progressed in 1939, there remained two macro-level tactical questions for the employment of the airborne forces in combat:

- How should paratroops attack a specific target, and
- How should the airborne forces seize an entire objective area?

The questions of where gliders and air-landing troops would arrive and how they would go into battle were self-evident. Gliders would land as close to their targets as possible, and air-landing troops would, by definition, land on secured airfields and be dispatched to the perimeter as required by the tactical situation. But how should paratroops be delivered relative to their targets? To land directly on prepared infantry positions would be suicidal. To land too far away without tactical vehicles and launch a conventional infantry attack would forfeit the element of surprise and put the lightly armed paratroops at a disadvantage against an enemy with artillery in prepared positions. Both approaches were debated. Ultimately Student decided the best tactic

---

[53] Edwards, pp 24-25. Radios and medical equipment were also dropped in the canisters.
[54] Ibid., p. 20.
[55] Interview with Sergeant Ernst Simon, *Fallschirmjäger Regiment* 1, Op cit.

would be to attempt to land as close to the target as possible and on as many sides of the target as possible simultaneously, preferably in a bell shape.[56] By nearly "surrounding" the enemy from the air, the enemy would be forced to defend everywhere at once.

The second and larger remaining doctrinal question was how the airborne assault itself should be structured. Should all the forces be concentrated in one place, then attack outwards? Or should a number of targets be attacked simultaneously by separate groups? To attempt the latter clearly violated the key principle of concentration of force. Student, however, strongly believed in this approach, which he described with the analogy of "drops of oil."[57] Groups of parachutists would land in several places, and, like oil drops, spread outwards until they merged together. With several battlefield options, the commander could reinforce the successful attacks, and, by not putting all his eggs in one basket, theoretically improve his chances of a success at some point. In this, Student was guided by Napoleon's maxim *"on s'engage partout et puis on voit"* ("One engages the enemy everywhere, then decides what to do.").[58] He was far from unopposed in this view, but as commander of the airborne forces he prevailed. Other officers, including senior men in his own command, correctly disagreed, among them Ramcke, Meindl and von der Heydte. It was to be Student's greatest blunder of the war.

Part V: Testing the Doctrine: Early Airborne Operations, 1939-1940

After the conquest of Poland, Hitler moved to secure his northern flank, using the *Fallschirmjäger* for the first time in the invasions of Norway, Denmark and the Low Countries.[59] Germany's first airborne operations succeeded beyond even Student's expectations. The first

---

[56] Department of the Army *Historical Study No. 20-232*, p. 54 (Appendix by von der Heydte)
[57] Ibid., p. 5.
[58] Department of the Army *Historical Study No. 20-232*, p. 5.
[59] Two parachute operations considered for the invasion of Poland were cancelled because of the speed of the German advance. Several companies of the Parachute Regiment did participate in patrolling and mopping up operations on the ground. The first of some 60,000 personnel assigned to German parachute units to be killed in action in World War II fell in Poland.

German parachute drop of the war took place on April 9, 1940.[60] One company of paratroops from *I. Bataillon, Fallschirmjäger Regiment 1* dropped on the bridge linking the Danish islands of Falster and Fyn. The Danes surrendered without opposition and German armored forces were soon rolling across the bridge. Two other companies from *I./FJR1* attempted to drop on the Oslo-Fornebu airfield in Norway but were aborted due to fog on the drop zone. On April 17, the remaining company of *I./FJR1* dropped on a road junction in the Gudbrandsdal valley near Dombas, Norway in an attempted blocking action. Of the 15 Ju52 transports carrying the company of 160 men to the drop zone, one was destroyed by anti-aircraft fire in the air and eight others flew off course. Just six reached the correct drop zone, and 61 paratroops were mustered on the drop zone under sporadic Norwegian fire. They managed to hold the road junction for four days until the surviving 34 men ran out of ammunition and surrendered. In May, the *I. Bataillon* of *FJR1* was deployed again to reinforce General Dietl's beleaguered garrison at Narvik but had little effect on the battlefield, which the British and French forces decamped in June.[61]

Valuable insights clearly presented themselves, but tragically for the German airborne, the most important one was lost. The Falster-Fyn bridge operation showed the tactical potential of the airborne as part of a combined arms offensive and the value of surprise in paratroop operations against light opposition. The aborted Oslo-Forneblu operation illustrated the extent to which weather impacted jump operations. The airborne reinforcement of Narvik demonstrated that paratroops could reinforce a garrison even when sea and land lines of communication were cut, a use to which paratroops have often been put in the years since. The drop at Gudbrandsdal, however, showed the consequences of dropping on a defended drop zone without fire support and aerial re-supply. The implications of this experience, with its 50% casualty rate foreshadowing almost exactly the fateful invasion of Crete a year later, were missed.

The greatest successes came in early May in the Battle of France. An assault team in gliders landed atop the Belgian fort of Eben Emäl, which the Belgian military believed to be impregnable. 77 German paratroop engineers armed with a new innovation - - shaped explosive charges - - succeeded in landing on the unmanned roof of the fort, and in short order put the entire gun system out of action and captured nearly 800 defenders.[62] Additional parachute groups captured bridges, airfields and road junctions across Belgium and Holland and opened the way for German armored columns to drive rapidly through the Low Countries and into France, forming the right hook of *Fall Gelb*.[63] The capture of the important Willems Bridge over the Maas River in the center of Rotterdam in particular demonstrated Student's remarkable flair for military innovation. Paratroops aboard a dozen old Heinkel 59 seaplanes landed on both sides of the river, taxied to the bridge abutments, jumped out at the water's edge, and secured them.[64] The element of surprise was often paralyzing for the Belgian and Dutch defenders, and overall casualties for the *Fallschirmjäger* were relatively light. General Student himself, however, was gravely wounded again in his second war, struck in the head by a sniper's bullet during the fighting in Rotterdam.[65]

These parachute actions stunned the western military world. Great Britain and the United States were finally spurred into action, rapidly creating parachute forces of their own. Even the U.S. Marine Corps got into the act, as the German army had in 1937, issuing an order on May 14, 1940 to

---

[60] The Soviet Union is credited with history's first combat parachute jump for a minor effort in Finland the previous year.
[61] Von Hove, pp. 47-55.
[62] Mrazek, James E. *The Fall of Eben Emael*. Self-published. Library of Congress no. 74-110765. p.183.
[63] Griess, pp. 45-46.
[64] Mason, Herbert Molloy, Jr. *The Rise of the Luftwaffe 1918-1940*. The Dial Press, New York. 1973. pp. 346-347
[65] Barnett, Correlli, ed. p. 470. Student's recovery took months. He fully regained his faculties, but spoke slowly afterwards, which led some to believe he had been made slow-witted as a result of his head wound. He was not.

begin forming a unit of parachute troops.[66] On the basis of German operations in the Low Countries and later the invasion of Crete, the Allies invested large amounts of men and materiel in the development of airborne forces during World War II. In the meantime, the successes in the Low Countries made instant national heroes of the *Fallschirmjäger*. Newsreels in Germany were full of their exploits;[67] magazines, books and even children's games were produced to celebrate their triumphs.[68] For General Student, the result was vindication of his principles and glory within the *Wehrmacht*. As a result, the airborne branch was ordered to be expanded rapidly, and adventurous volunteers from all across Germany and throughout the military flocked to the new force.

Part V: 1941: The Turning Point: Crete or Malta?

After the conquest of France, Hitler had little interest in his war with Great Britain. Calling off the Battle of Britain on the eve of victory over the RAF,[69] he turned his attention to the East. Hitler was in fact obsessed by his desire to destroy the Soviet Union and the plans for *Operation Barbarossa* to implement it.[70] To his annoyance,[71] he had been recently forced into a campaign through the Balkans into Greece and across the Mediterranean into North Africa by Mussolini's comic opera military and a string of Italian military disasters.[72] Already, a small corps of two divisions and support troops under General Erwin Rommel had been dispatched to North Africa,[73] and German forces had rolled through Yugoslavia. Now, at the *Führerhauptquartier*, Hitler, Field Marshall Keitel and General Jodl were discussing the end of the Balkans Campaign. Should German forces

---

[66] The Marine parachute forces eventually grew to regimental size and fought with valor on Guadalcanal and Bougainville. Because of unsuitable terrain throughout most of the Pacific theatre, the Marine paratroops never made a combat jump and were disbanded in 1944. Most formed the nucleus of the 27th Marine Regiment, which fought on Iwo Jima and seized Mount Suribachi. Two of the Marines who participated in the second flag-raising on Mount Suribachi, immortalized by the Joe Levanthal photograph and the sculpture at the Marine Corps Memorial in Arlington, were former "paramarines." Hoffman, LTC Jon. *Silk Chutes and Hard Fighting: U.S. Marine Corps Parachute Units in World War II*. History and Museums Division, Headquarters Marine Corps. Washington, DC. 1999.
[67] Hickey, p. 56.
[68] Kurtz, Robert. *German Paratroops*. Schiffer Publishing. Atglen, PA. 2000. p. 184.
[69] Magenheimer, p.35.
[70] Bullock, Alan. *Hitler and Stalin*. Vintage Books, Random House. New York. 1991. pp. 687-688.
[71] Von Hove, p. 111.
[72] Magenheimer, p. 69.

end their southern thrust by capturing Malta, or by taking Crete? Keitel and Jodl advocated an attack on Malta.[74] British RAF fighters based on Malta were already impacting the maritime re-supply of the fledgling *Afrika Korps*.[75] The *Kriegsmarine* also urged that Malta be taken to safeguard the supply convoys to Rommel. Fully aware that the Royal Navy's dominance in the region[76] precluded any amphibious invasions, Hitler had had the *Fallschirmjäger* in mind for both options as early as December 13, 1940.[77] In April, 1941, he summoned *Reichsmarshall* Göring, General Student, and General Jeschonnek, formerly Student's Chief of Staff and now Chief of Staff of the *Luftwaffe*, to his headquarters.[78]

Earlier in the year, Student had submitted a plan for an airborne attack on Crete. Now, on April 21st at the strategy conference in Hitler's inner sanctum to decide the question of Crete or Malta, he was asked to present his proposal. Student, still a relatively junior air force general, advocated Crete as the target of his paratroops, arguing that the island could be used as a springboard for further operations into the Middle East and Iraq after *Operation Barbarossa* was concluded. Keitel and Jodl again made the case for Malta, but Göring[79] and *Luftwaffe* General Löhr[80] supported Student. Then Student played his trump card: There was a threat to the oil supply for *Barbarossa*, a critical vulnerability of the Nazi war machine. British heavy bombers, operating from improved airfields on

---

[73] Bender, R. James and Law, Richard. *Afrikakorps*. R. James Bender Pub. San Jose. 1973. pp. 22-25.

[74] Student and Götzel, pp. 198-199.

[75] Barnett, p. 301. Rommel was not advised of *Operation Merkur* until it was over. When informed of the operation afterwards, he remarked in disgust: "Malta! Malta, not Crete, is the key to the Mediterranean!" Fraser, David. *Knight's Cross, A Life of Field Marshall Erwin Rommel*. Harper Collins Publishers. United Kingdom. 1993. p. 224.

[76] Reflecting Hitler's predilection for land war, Germany had gone to war in 1939 against the advice of his senior admirals with a very weak navy in comparison with Great Britain. Like Napoleon, Hitler had no appreciation of the importance of naval power. The senior German admiral at the end of the war, Karl Dönitz, commented after the German surrender : "Seldom has any branch of the armed forces of a country gone to war so poorly equipped." Admiral Erich Raeder once called Hitler "hopelessly *landsinnig* (land-minded)."

[77] MacDonald, p. 56.

[78] Kurowski, Franz. *Der Kampf um Kreta*. Maximillian Verlag. Herford. 1965. p. 9.

[79] In his memoirs, Keitel stressed Göring's support for the Crete option (Keitel, Wilhelm. *The Memoirs of Field Marshall Wilhelm Keitel*, ed. Walter Gorlitz. Cooper Square Press. New York. 2000. p. 142.). Historian Edwin Hoyt has suggested that Göring, ever mindful of his position and prestige in the Third Reich hierarchy and embarrassed by the failure of the *Luftwaffe* to win the "Battle of Britain," backed Student because he believed, as Student himself did, that Crete would be the softer target. (Hoyt, Edwin. *Hitler's War*. Plenum Publishing. New York. 1988. p. 176.)

[80] Department of the Army. *Pamphlet No. 20-260, The German Campaign in the Balkans (Spring 1941)*. November 1953 p. 120.

the island of Crete, would be just within operational range of the strategic German oil supplies at Ploesti, Romania.[81] By seizing Crete, Hitler could deny it to the RAF, and secure his oil supply from aerial interference. Indeed, ever since writing the polemical *Mein Kampf*, Hitler's strategic thinking had been focused on the Soviet Union and the "Bolshevik threat",[82] and despite his senior commanders' advice, he remained uninterested in the Mediterranean theatre beyond its impact on *Barbarossa*. He decided on Crete.[83] The attack was code-named *Operation Merkur* (Mercury),[84] and the pressure of time on Student to execute it was enormous: Hitler wanted the operation over as quickly as possible to get on with his war against Stalin.[85]

Part VI: "Pride Goeth Before Destruction"[86]: Crete 1941

For *Operation Merkur*, Student planned to repeat his Low Country tactical raid approach at the campaign level. He saw Crete in the same conceptual framework. Parachute and glider assault groups would capture tactical objectives, in this case the vital Cretan airfields, and reinforcements would be flown in immediately (rather than come overland as was the case in Holland and Belgium) to expand the airheads and overwhelm the British garrison. The paratroop assault groups would land like his "oil drops" to spread outwards from key objectives.[87] But in reality Crete would prove to be very different. The flat terrain of Holland and the rapid advance of the German main forces had glossed over the many inherent disadvantages of paratroops. Their short combat staying power, the dispersion of forces caused by the scattering of wind-blown parachutes, and their lack of tactical mobility once on the ground had not been critical in Holland. Only small numbers of parachutists had been required for each tactical objective, and each group was only required to hold its ground for

---

[81] Kiriakopoulos, G.C. *Ten Days to Destiny*. Franklin Waltz. New York. 1985. p. 38.
[82] Bullock, Alan. *Hitler and Stalin*. Vintage Books, Random House. New York. 1991. pp. 687-688.
[83] Farrar-Hockley, p. 88.
[84] Edwards, p. 81.
[85] Hickey, p. 60.
[86] Proverbs xvi 18
[87] Department of the Army. *Historical Study No. 20-232*. p. 5.

a day or two against light opposition within easy range of their own close air support.[88] On the flat ground of the Low Countries, assembly on the drop zones was relatively easy, and tactical mobility on that scale of objectives was not a major problem.

On Crete, however, the rugged landscape of steep rocky hillocks, sheer gullies, ravines, and shrubby olive groves would reduce the visibility of a standing man to an average of about 20 yards, less than the average dispersion of two parachutes,[89] and "wholly inadequate" reconnaissance failed to show the difficulty of the terrain, leading "both the command and the troops to erroneous conceptions."[90] Tactical air support would have to come from the Greek mainland, not nearby German airfields, and would have to use the same limited air control and refueling facilities, and the same extended supply lines, as the transports carrying the parachutists themselves.[91] Most critically, the objectives were not small tactical positions like bridges, which can be held at both ends by even a determined company of infantry, but whole airfields, with much larger perimeters and buffer zones against enemy supporting arms fire. Instead of companies or battalions, whole regiments would be needed, at a minimum, to capture and hold them.

Student blundered badly in employing the conceptual "oil drops" model of the successful tactical raids in the Low Countries in such a major operation. Both General Löhr at 4th Air Force, with

---

[88] Kurowski, *Deutsche Fallschirmjäger*. pp. 76-79.
[89] The island of Crete, like all of the Aegean, is essentially dry, with occasional downpours and flash floods. This has created a harsh landscape of erosion, marked by countless dry gulches and rocky slopes. What appears flat from the air is actually an infantryman's nightmare on the ground. It also provided superb natural camouflage for defensive positions, from the ground and from the air. (Observations derived from author's terrain walk, June 1986.) German parachutes opened by static line. The canopies deployed rapidly with a powerful shock, necessitating a head first diving exit to spread the opening shock evenly over the jumper's body. The average paratrooper fell 110 feet between exit and full canopy deployment, and then descended at 16 feet per second. The average time in the air for a jump from 400 feet was about 20 seconds, and given tight exit discipline, each jumper was approximately 16 feet below and 150 feet behind the next man to exit. If the three *JU52* transports held a tight formation, the 36 man platoon they carried would therefore be dispersed over an area 120 yards wide by 500 yards long. Price, Alfred. *Luftwaffe Handbook 1939-1945*. Charles Scribner's Sons, New York. 1977. p. 57.
[90] Department of the Army *Historical Study No. 20-232*, p.5.
[91] Kurowski, *Deutsche Fallschirmjäger*. pp. 152-153.

overall operational control of the three subordinate *Fliegerkorps* involved in Merkur,[92] and General *Freiherr* Wolfram von Richthofen,[93] commander of the close air support squadrons in *VIII Fliegerkorps*, saw this and proposed instead a concentrated assault on the western end of the island, followed by an eastern thrust to capture the key objectives. But Student, controlling the paratroops of *XI Fliegerkorps,* feared RAF use of the eastern airfields for interdiction and difficulties with the mountainous terrain in the west. Flush with victory in Holland and Belgium, Student insisted on his oil drops approach, proposing seven separate drop zones and objectives. A compromise proposed by the *Oberkommando der Luftwaffe* (OKL) and approved by Göring was adopted that reduced the number of drop zones but maintained the flawed approach.[94] The *Luftwaffe* would split its forces and attempt to seize four separate objectives, the three island airfields and the best port facility, simultaneously.[95] Not unlike Lee at Gettysburg, an outstanding commander had come up with a truly awful plan, and failed to listen to the advice of competent officers who knew it. Student, with occasional lapses like the inefficient weapons delivery method, had done a brilliant job of developing an entirely new and untried method of warfare in just three years. Yet this one singular mistake would lead directly to the fall from grace of his beloved paratroops.

Student's faulty application of raid tactics to his concept of a full-blown airborne assault already endangered the entire operation. The plan now ran headlong into the other three horsemen of the military apocalypse: poor intelligence, bad equipment, and dismal security. His intelligence officer, *Major* Reinhardt produced what could very well be one of the worst intelligence

---

[92] *IV Fliegerkorps* – transport and supply, *VIII Fliegerkorps* – close air support, *XI Fliegerkorps* – the paratroop force.
[93] A cousin of the "Red Baron" Manfred von Richthofen of the First World War. The Richthofen family had, and still has, a long and broad involvement in the German Armed Forces.
[94] Farrar-Hockley, pp. 90-91.
[95] Kuhn, p. 57.

appreciations of all time: He estimated that as few as 5,000 defenders were on the island,[96] and believed these to be poorly armed and second-rate forces.[97] In fact, there were more than 27,500 well-armed, first rate Commonwealth troops holding the island with good artillery and 24 light and medium tanks.[98] In addition, some of the best forces of the Greek army, which had routed Mussolini's forces the year before,[99] were on the island, numbering a further 14,000 men.[100] Altogether, the 8,300 German paratroops were facing 41,500 defenders. Once they had all landed, the attackers would be outnumbered by the defenders exactly five to one.

A second weakness lay in the use of the separate arms containers. On the corrugated, rocky landscape of Crete, these containers would prove to be a devastating flaw in the plan. Many weapons canisters were never found, others would be dropped into Commonwealth positions or directly under their guns. This not only delayed attacks and drastically reduced the paratroopers' firepower and all-important "shock effect," it also resulted in hundreds of casualties on Crete among unarmed paratroopers desperately searching for something with which to defend themselves.[101]

Finally, there was the failure of security. On August 16, 1939, Polish intelligence had given the British a working German Enigma coding machine together with a complete set of the critical Zygalski code sheets[102], and a dedicated group of cryptanalysts at Blechley Park had done what the *Abwehr* believed to be impossible: They had broken the German operational level codes.[103] The secret decoding operation, called "Ultra," in this case provided the senior British officer on the

---

[96] Beevor, p. 79.
[97] MacDonald, p. 80. Ironically, Montgomery would make exactly the same mistake in planning for Operation Market Garden three years later.
[98] Ibid., p. 32.
[99] Ibid., p. 52.
[100] Department of the Army. *Pamphlet No. 20-260.* p. 123
[101] MacDonald, p. 175.
[102] Freedman, Maurice. *Unravelling Enigma.* Pen & Sword Books, Ltd. South Yorkshire, England. 2000. pp. 21-22.

island, General Bernhard Freyberg, VC, with complete knowledge of where, when and in what force the Germans would arrive, and what they planned to do when they got there.[104] Although Freyberg could not disclose the source or substance of the information to his staff, he could make his dispositions accordingly. The stage was set for a total military catastrophe for the Germans.

Part VIII: The Battle Plan

By May 1941, after exactly five years of development, Germany still only had four regiments of trained parachutists available, *Fallschirmjäger Regiments 1, 2* and *3*, and the elite Assault Regiment, the *Luftlande Sturm Regiment*, many of whose paratroopers had also been trained in gliders. Each regiment would essentially be assigned one of the four objectives. There were not enough transport aircraft available to bring all the assault troops in the first wave, so the transports would have to drop the first wave, return to southern Greece, refuel, load the second assault wave and their weapons canisters, and make their way back across the Aegean to drop them.[105]

The final operational plan called for the reinforced *Sturm Regiment* with about 2,300 men under Major General Eugen "Papa" Meindl[106] to attack and seize the airfield near the village of Maleme in the first wave on the morning of May 20th. Six hundred of them would arrive in gliders. In the other prong of the first wave, *Fallschirmjäger Regiment 3,* with about 2,000 men under *Oberst* (Colonel) Heidrich, was ordered to take the port town of Cania and block a key road junction. In the second wave, *Fallschirmjäger Regiment 2* with about 2,000 paratroops under *Oberst* Sturm

---

[103] MacDonald, p. 136.

[104] Bennett, Ralph. *Ultra and Mediterranean Strategy.* William Morrow & Company, New York. 1989, pp 51-62.
[105] This has been attributed in most Allied sources to logistics and air traffic control bottlenecks in southern Greece (Stewart, I.McD.G. *The Struggle for Crete*. Oxford University Press. London. 1966. p. 87 et al), but this explanation is incorrect. The actual reason was insufficient transport to bring all the paratroops to the island at the same time. Student had at his disposal somewhat fewer than 500 operational Ju52 transports, each of which was capable of carrying 13 paratroopers. (See Kurowski, *Der Kampf um Kreta*, p. 22, Busch, p. 33, Murray and Millet, *A War to be Won*, Harvard University Press. Cambridge, MA 2000, p. 106).

would seize the minor port and airfield at Rethymnon. At the same time, *Fallschirmjäger Regiment 1,* with an additional 2,000 paratroops under *Oberst* Bruno Bräuer was to take the island's main airfield at Heraklion together with the town itself (see map, Appendix A).[107] The four individual attack plans were left to the discretion of the regimental commanders and their staffs.[108] Overall command on the ground on Crete was given to Major General Wilhelm Süssman.[109]

The *Luftwaffe* would bring the paratroops and gliders to battle, provide aerial re-supply of ammunition, food and medical supplies to the paratroops on the ground, and bring reinforcements to the battle as directed by General Student in Athens. Without significant integral fire support assets, the paratroops would be almost entirely dependent on tactical air support from the *Luftwaffe*.[110] Fighter-bombers would precede the transports to engage targets around the objectives, while others would remain on station for air strikes called by liaison officers with radio sets attached to each battalion.[111] Aerial recognition panels would be used to mark friendly lines. As soon as an airfield was secured, the *Luftwaffe* would land reinforcing infantry. The 22nd Air Landing Division was unavailable, so the 5th Mountain Infantry Division under General Julius Ringel was earmarked for the operation.[112] Reading the battle plan for the first time shortly before the operation, Ringel wrote after the war, "sent a cold chill down one's spine, for it was clear that the operation…would be a suicidal adventure."[113]

---

[106] Short, scrappy and extremely well liked by his troops, at age 49, Meindl was one of the oldest paratroopers in the German armed forces. Major General was the lowest flag rank in the *Luftwaffe*, as there were no brigadier generals in the *Wehrmacht*.

[107] Objectives: Kuhn, p. 62. Unit strengths: Stewart, p. 85. German sources sometimes also refer to the town of Heraklion by the German spelling, "Iraklion," which is used in a number of texts and is closer to the Greek pronunciation of the name.

[108] Kurowski, *Der Kampf um Kreta,* p. 13.

[109] Beevor, p.104.

[110] The *Luftwaffe* did experiment with dropping light artillery during the campaign using a bundle of five parachutes together, called a *Fünfling*. The guns were almost always damaged, however, and significant quantities of ammunition could not be provided. The experiments were considered unsatisfactory and remanded to the *Lehr* (Test) Battalion formed after the battle for further study, together with solving the problem of jumping with weapons.

[111] Stewart, p. 88.

[112] Lucas, p. 46.

[113] Kuhn, p. 64.

At Hitler's insistence, so that, as he put it, they would not be "standing on one leg," additional German forces were loaded onto a motley collection of small craft for sea-borne reinforcement. Operating from the Greek port of Piraeus, these troops would sail to the small island of Milos, then transit the dangerous 70 miles of open sea between Milos and Crete in two waves during daylight on May 21st and 22nd, relying on *Luftwaffe* air superiority[114] to protect them from the Royal Navy.[115] Both Ringel and Student had serious reservations about this, but incredible as it seems, there simply was no planning for their amphibious landing on Crete beyond sailing for the port of Cania, which, it was assumed, would be in German hands when they arrived.[116] The top speed of the vessels was four knots, or about five miles per hour.[117] One Italian corvette and several Italian motor torpedo boats were assigned to each wave as escorts.[118]

Part IX: Pyrrhic Victory

Following a partially successful attempt to trap retreating British forces in Greece with a parachute jump at the Corinth canal by the 2nd Parachute Regiment in early May,[119] the German paratroop division settled into bivouac near the few airfields in southern Greece.[120] In the pre-dawn hours of May 20th, the regiments in the first wave boarded the Ju52 transports. Even before the first paratrooper jumped over Crete, however, the assault plan began to come unglued. The first casualty

---

[114] MacDonald, p. 72.
[115] Ibid. p. 61.
[116] Stewart, p. 90.
[117] Ibid., p. 280.
[118] Ibid. Even this limited Italian assistance was reluctant, and the Germans put no faith in the participation of the Italian navy.
[119] Lucas, pp. 38-45.
[120] Kurowski, Franz. *Deutsche Fallschirmjäger 1939-1945*. Edition Aktuell, Frankfurt, c. 1990, pp. 132-138.

of *Operation Merkur* was the commander of the attacking forces, General Süssman, who was killed together with his twelve senior divisional staff officers in a freak glider accident over the Aegean.[121]

As soon as the transports reached the island, the low-flying, lumbering Ju52s became easy targets for anti-aircraft fire and a score were destroyed in the air. Many paratroops were released directly over British positions, and hundreds were killed in the air. Hundreds more were killed immediately upon landing, some by Cretan partisans. The attack in many places resembled a massacre. The 480 men of 2nd Battalion, *Fallschirmjäger Regiment 1*, for example, jumped directly on top of the Argyll and Southerland Highlanders, and suffered 312 men killed and 108 men wounded in 20 minutes without getting off the drop zone.[122] Heavy flak and the confusion of war throughout the day resulted in many of the attackers being widely scattered and often landing far from their objectives.

The attacks on Cania, Rethymnon and Heraklion were disasters. Expecting an under-strength battalion in Heraklion, for example, *Fallschirmjäger Regiment 1* ran into a full British regiment in the town. They made headway in street fighting until nightfall when they finally ran out of ammunition and withdrew from the city.[123] *Fallschirmjäger Regiment 2* at Rethymnon and *Fallschirmjäger Regiment 3* at Cania fared no better. By nightfall, the survivors from all three groups were surrounded and rapidly running out of the means to resist. They radioed their situations to Student's command center in Athens. The waterborne reinforcements had their own version of Hell. The Royal Navy, with intelligence provided by Ultra, used the cover of night to find

---

[121] It was established after the battle that Süssmann's tow pilot emerged from cloud cover to find himself on a collision course with another JU52 transport. He apparently dived to get under the other aircraft, but some combination of possible structural defect, airframe stress and turbulence caused the glider's left wing to shear off. The tow rope parted, and the glider fell a thousand feet, killing Süssmann and most of the division's senior staff officers, twelve of whom were traveling with him at the time. Neither Student in Athens nor the senior surviving officers on Crete were immediately aware of what had happened. Kiriakopoulos, p. 131.
[122] Stewart, p 206.
[123] Account of LTC Wilhelm Reinhardt, *Fallschirmjäger Regiment 1*. Interview with the author, Bruchsal, Germany 1996.

the slow-moving wooden transports of the first wave, which were still far from the coast of Crete by nightfall, and annihilated them.[124] The second amphibious wave was aborted.

At the end of the first 12 hours of battle, almost half the paratroopers in the German armed forces, over 3,000 men, were dead or wounded. Most of the rest were surrounded by superior forces and in danger of destruction in detail.[125] Only at Maleme did the *Sturm Regiment* still have any significant freedom of maneuver.[126] The steady New Zealand defenders there remained firmly in control of the high ground, however, particularly that of Hill 107, where the 22nd N. Z. Infantry Battalion under LTC L.W. Andrew, VC, completely dominated the airfield.[127] Of the 2,300 German attackers in the *Sturm Regiment*'s morning assault, less than 600 remained unwounded in two pockets about a mile apart[128] at nightfall, sheltered in the gullies at the foot of Hill 107.[129] Many were heat casualties after fighting all day in woolen uniforms[130] in 100°F heat without water. Ammunition supplies were nearly exhausted.[131] The regimental commander, Meindl, had been shot through the chest and was believed to be dying. Virtually all the company officers and two of the four battalion commanders were dead or critically wounded.[132] A third, *Hauptmann* (Captain) Gericke, was also wounded but now commanded the regiment. Radio communications from the *Sturm Regiment* were intermittent,

---

[124] Of the 2,331 soldiers in the convoy, 325 were killed, 36 men struggled ashore at the western tip of Crete, and the remainder were rescued by *Luftwaffe* air-sea rescue units and Italian motor torpedo boats and brought back to Greece. Stewart, p. 280.

[125] Stewart, p. 229.

[126] Of the four attacks, the one only still even marginally viable at this point was the one which employed a significant number of gliders in the assault. The concentration of gliders at Maleme was determined after the battle to be influential in establishing such a foothold as the Germans were able to gain there. The glider missions there were rated by the Germans after the battle as 75% accurate. Headquarters, United States Army Air Forces in Europe (Office of the Assistant Chief of Staff, A-2) Report of October 15, 1945 *Air Staff Post Hostilities Intelligence Requirements on German Air Force: Tactical Employment Troop Carrier Operations*. p. 195.

[127] Kiriakopoulos, p. 236-237.

[128] Stewart, p. 241.

[129] Dr. Fink, Feri. *Der Komet auf Kreta*. Gelka-Druck und Verlags. Ettlingen. Undated c. 1990. p.272.

[130] Tropical uniforms were ordered but did not arrive in time for the operation, apparently due to a quartermaster error. *Air Staff Hostilities Intelligence Requirements on German Air Force*, p. 185.

[131] German Army Research Report, *Der Luftlandeangriff auf Kreta*. Dokumentationszentrum der Bundeswehr, Bonn, Film Nr. 984, p. 45.

[132] Beevor, p. 151. *III. Bataillon* of the *Sturm Regiment*, for example, began the day on May 20 with 12 officers and by nightfall 10 of them were killed or wounded. Only two junior medical officers survived the battle unharmed. Winterstein and Jokobs, *General Meindl und Seine Fallschirmjäger*. Verlagsbuchbinderei Ladstetter GmbH. Hamburg. 2. 1990, p. 276.

and Student had no clear picture of the critical situation at Maleme.[133] The British were within hours of victory, and the Germans were on the brink of a catastrophe.

British victory was not to be. Within 10 days, the Germans had salvaged a costly victory in spite of the overwhelming odds. Armed with foreknowledge of the entire German plan, Freyberg chose to disregard it, even after capturing a copy of the German operational plan on the first day of battle and having the attacks of May 20th confirm the information in detail. He continued to expect the "main effort" to come by amphibious assault, and maintained many of his best troops by the water's edge.[134] Believing his position untenable, Colonel Andrew radioed for permission to withdraw from General Hargest, commanding the New Zealand brigade. Hargest, miles from the front and entirely out of touch with the tactical picture, responded: "If you must, you must."[135] Andrew ordered his 22nd N.Z. battalion off Hill 107 during the night of May 20-21.[136]

During the night, German patrols were astonished to discover the New Zealanders had pulled out. Just before dawn on the 21st, one of the few surviving *Sturm Regiment* officers, Dr. Neumann, the regimental surgeon, led the survivors up the hill and into the vacated positions.[137] They marked the trenches with recognition panels so their own fighters wouldn't attack them at first light.[138] At almost the same time, General Student dispatched a staff pilot in a JU52 from Athens to determine the situation at Maleme.[139] With no fire from Hill 107 to prevent it, the pilot, a *Hauptmann* Kleye, succeeded in landing and taking off from the airstrip. With the aggressive instincts of a fighter pilot,

---

[133] Haupt, Werner. *Fallschirmjäger 1939-1945, Weg und Schicksal einer Truppe*. Podzun-Pallas Verlag. Friedberg. 1979. p. 61. Due to equipment losses, the *Sturm Regiment* was reduced to Morse code communications.

[134] Freyberg has been described as "obstinate and at times almost willfully obtuse." MacDonald, p. 142.
[135] Time-Life Books, ed. *The Conquest of the Balkans*. Time-Life Books, Inc. Alexandria, VA. 1990. p. 130.
[136] MacDonald, p. 200.
[137] Neumann was awarded the Knights Cross for his leadership on Crete. Thomas and Wegmann. *Die Ritterkreuzträger der Deutschen Wehrmacht 1939-1945, Teil II: Fallschirmjäger*. Biblioverlag. Osnabrück. 1986. p. 196.

Student gambled on sending in his only *ad hoc* reserves, two battalions of paratroops left behind from the second wave the day before. About 550 men under *Oberst* Ramcke dropped west of Hill 107 and reinforced the airfield.[140]

In the late afternoon of the 21st, although the airstrip at Maleme was still under indirect British artillery fire, the first JU-52 transports bearing a battalion of Army infantry reinforcements from Ringel's 5th Mountain Division managed to land and deliver troops, ammunition, and supplies.[141] By nightfall, however, the Germans still had fewer than 1,800 men in the Maleme sector fit for action.[142] Although he now outnumbered the attackers there 20:1, to the exasperation and anguish of British junior officers, General Freyberg failed to mount a timely and aggressive counter-attack.[143] When he did finally attempt a counter-attack, it was late, weak and tentative.[144] Student, as aggressive and decisive as Freyberg was timid and remote from the battlefield, drove rapidly to widen the wedge in the island's defenses. Because the airstrip was so short, pilots had to land into the wind on the short runway, then turn, taxi the whole length of the field, and take off again into the wind, all while under artillery fire. Transports were hit regularly, until the airfield was littered with wrecks, but German mountain troops poured onto the island on the 22nd.[145] Although they were never outnumbered less than 3:1, the remaining paratroops and the mountain troopers attacked aggressively from the expanding airhead, linked up with scattered the parachute pockets, and drove the Maleme and Cania defenders to the small port of Sphakia on the southern coast. Several

---

[138] Haupt, p. 61.
[139] Kurowoski, *Der Kampf um Kreta.* p. 86-87.

[140] MacDonald, p. 196. Ramcke command and number of men, see Kuhn p. 104. Ramcke later led a brigade in North Africa and commanded the German garrison at Brest in 1944.
[141] Kuhn, pp. 107-108.
[142] Veranow, p. 186.
[143] Kiriakopoulos, p. 224.
[144] MacDonald, pp. 211-212.
[145] Mitcham, *Hitler's Legions,* p. 338.

thousand troops were evacuated from Heraklion and Sphakia by the Royal Navy. By June 1st, the battle was over: Freyberg, Hargest and Andrew had snatched defeat from the jaws of victory.

Part X: The Aftermath of Crete: Two Year Hiatus, 1941-1943

The Germans lost 4,054 men killed in action on Crete, including 3,119 paratroopers, the heart of the airborne.[146] Another 2,800 Germans were wounded, many of them paratroopers too seriously injured to return to active duty. The *Luftwaffe* lost 350 aircraft, at least half of them Ju52 transports sorely needed for *Operation Barbarossa*.[147] Crete remained in German hands throughout the war, and was in fact the last German garrison to go into captivity in the summer of 1945,[148] but strategically its capture accomplished little. Within a year, the first bomber attack on Ploesti was conducted by the "Halpro Group" from an air base at Fayid, Egypt.[149] For this brief respite, the Germans paid a terrible price. Although German casualties on Crete would pale in comparison to losses suffered on the Russian Front, they decimated the German parachute force, and it would be a year before their ranks were restored to their pre-*Merkur* levels.

Student was deeply affected by his losses on Crete. To one of his battalion commanders, Friedrich von der Heydte, he appeared after the battle a broken man. "He had visibly altered. He seemed much graver, more reserved, and older. There was no evidence in his features that he was joyful over the victory - his victory - [nor] proud of the success of his daring scheme," von der Heydte wrote after the war, "the cost of victory had evidently proved too much for him."[150] Hitler, for his

---

[146] Kurowski, *Der Kampf um Kreta*, p. 236.
[147] *Department of the Army Pamphlet 20-260*, p. 141. Ringel estimated the number of transports lost at 200, while Halder gave a figure of 170 transports lost on May 28. (Stewart p. 476).
[148] Kiriakopoulos, p. 369.
[149] The "Halverson Project No. 63," or Halpro Group, made the 2,600 mile round trip mission with 13 B-24 Liberators on June 11, 1942 under the command of Colonel Harry Halverson. "Operation Tidal Wave" inaugurated bombing missions from Benghazi on August 1, 1943, which became known as "Black Sunday" for the losses to the 376th, 93rd, 44th, 389th and 98th Bomb Groups over the target. Michael Hill, *Black Sunday, Ploesti*, Schiffer Publishing, Atglen, PA, 1993. p 11.
[150] Von der Heydte. *Daedalus Returned*. Hutchinson Publishing. London. 1958. p. 180. General Prof. Dr. Friedrich August *Freiherr* von der Heydte was one of the most remarkable German paratroop officers of the Second World War. He commanded the *I. Bataillon* of *FJR3* on Crete, where he buried his own dead side by side with the Commonwealth and Greek casualties in a

part, was appalled by the casualties,[151] which were the highest for any German operation to date. They were never released to the German public. For the Nazi propaganda machine, Crete was trumpeted as another glorious triumph of German arms, and it is difficult to say whether the German people or Allied airborne planners were more greatly deceived by this canard.

The truth was far different. Student's "oil drops" approach had indeed been catastrophic, and the military organization that had literally taken him years to build was in ruins. Hitler drew the wrong conclusions from the disaster. In his mind it was not a failure of doctrine or tactics that was responsible, but a lack of surprise. At an awards ceremony for the survivors of Crete, he pulled Student aside and informed him: "Crete has shown that the days of the paratroops are over. Paratroops are a weapon of surprise, and the surprise factor has been overplayed."[152]

What had actually happened? Student wrote after the war that the very name of Crete "conjured up bitter memories. I miscalculated when I proposed the operation, and my mistakes caused not only the loss of very many paratroops, whom I looked upon as my sons, but in the long run led to the demise of the German airborne arm I had created."[153] Clearly Löhr and Richthofen had been right: Student's tactical approach meant not success but weakness everywhere.

---

common memorial. He served in North Africa and commanded *FJR6* in Normandy in 1944, earning the nickname "the Lion of Carentan." Including Holland and the Ardennes operation which he commanded in December 1944, he made three combat jumps in WWII. Von der Heydte was implicated in the Hitler assassination attempt of July 1944 and escaped execution by a typographical error on the part of the Gestapo. After the war he wrote a book on irregular warfare entitled *Modern Irregular Warfare in Defense Policy* in 1972 in addition to his memoir of the battle of Crete, *Daedalus Returned*, in 1958. He was a General of the *Bundeswehr* Reserves, a Professor of International Law at the University of Mainz and served in the Bavarian State Parliament.

[151] Hart, Liddel. *History of the Second World War*. Da Capo Press. New York. 1999. p. 138.
[152] Student and Götzel, p. 337.
[153] Alman, Karl. *Sprung in die Hölle*. Erich Pabel Verlag GmbH & Co. Rastatt. 1964. p.273. Student's own son was killed in action flying in the *Luftwaffe* in 1944.

Student recognized this, but attributed his decision to faulty intelligence. During post-war interrogations, Student stated that "if he had had correct information on the enemy disposition and strength, he would not have dispersed his parachutists over a 75-mile front, but would have concentrated all his parachutists around Maleme and captured the airdrome the first day and air landed the ground troops the second day, then massing [sic] forces to take the island."[154] Certainly, Student and his staff had underestimated both enemy strength and the problems of delivering parachutists to battle, errors British Field Marshall Montgomery would ironically repeat precisely in Operation Market Garden in 1944 when Student commanded the German defense on the ground in Holland. The sequence of "gliders-paratroops-reinforcements" was not wrong, but the paratroops had to be landed *en masse* with their weapons and safely out of range of enemy direct fire and artillery observation, and on Crete they were not.

After the war, in a feud reminiscent of the post-Civil War animosity between Pickett and Lee, the *Sturm Regiment*'s commander, General Meindl, was sharply critical of Student's decision, writing that Student "had big ideas but not the faintest conception as to how they were to be carried out."[155] Like Pickett, Meindl's command had been virtually wiped out as a result of the battle. General von der Heydte was more circumspect. In 1958, von der Heydte wrote of his experience listening to Student brief his officers in Athens 17 years before: "[*Operation Merkur*] was his own personal plan. He had devised it, had struggled against heavy opposition for its acceptance, and had worked out all the details. One could perceive that this plan had become a part of him, a part of his life. He

---

[154] Headquarters, United States Army Air Forces in Europe (Office of the Assistant Chief of Staff, A-2) Report of October 15, 1945 *Air Staff Post Hostilities Intelligence Requirements on German Air Force: Tactical Employment Troop Carrier Operations.* p. 196.
[155] Shulman, Milton. *Defeat in the West.* Secker and Warburg, London. 1947. p.58.

believed in it and lived for it and in it."[156] From the remove of 60 years after the battle, the most likely assessment of what happened in April 1941 may be that Student, in a rare but terrible lapse of judgment, fell in love with his own plan and his own tactics. It is a danger for any commander in any war.

The outcome is not in dispute: The German airborne committed suicide as a parachute force in ten days in May 1941. After the war, Student called Crete "the graveyard of the German paratroops."[157]

Student feared at first the *Fallschirmtruppe* might be disbanded by Hitler altogether.[158] Their scattering over the Russian Front that winter as conventional infantry, out of German desperation to stabilize the situation there, reflected Hitler's thinking in his brief meeting with Student after Crete. But Hitler, either by oversight or distraction with his crusade against the Soviet Union, never formalized his decision to disband them, and the German airborne arm actually continued to expand.[159] After Crete, additional jump schools were opened in Germany, training output of parachutists gradually quadrupled, recruiting standards were relaxed somewhat, and a steady flow of volunteers gradually filled out the 1st Parachute Division with trained paratroops despite their heavy losses on the Russian Front. Jump schools now taught trainees to jump with their weapons.[160] New arms were developed, including the more potent 4.2 cm PAK to succeed the 3.7 cm version,[161]

---

[156] Von der Heydte, p. 40.

[157] Interrogation of *Generaloberst* Student, 27 September 1945. Summarized in Headquarters, United States Army Air Forces in Europe (Office of the Assistant Chief of Staff, A-2) Report of October 15, 1945 *Air Staff Post Hostilities Intelligence Requirements on German Air Force: Tactical Employment Troop Carrier Operations.* p. 193.
[158] Farrar-Hockley, p. 103.
[159] Ibid, p. 104.
[160] British Intelligence M.I.14 Report M.I.14(j)/su/28/44 "*Training and Tactics of Parachutists, Winter 1943/44,*" National Archives microfiche 512.6312-29, declassified. pp. 1-11.
[161] The 3.7 cm version of the gun was described by one veteran of the *Fallschirmtruppe* as "very ineffective." (interview with Sergeant Ernst Simon, Op cit.)

10cm LG2 recoilless guns, and the revolutionary *Fallschirmjäger Gewehr 42* assault rifle,[162] which fired single rounds from a closed bolt and bursts from an open bolt. It was shorter than the K98, weighed just 10 pounds, and fired standard rifle ammunition from a 20-round magazine.[163]

At Rommel's urging, Hitler began to see the need to capture Malta. He agreed in principle to an airborne invasion of the island. Planning for an assault on Malta in 1942, called *Operation Herkules*, proceeded seriously enough to include the movement of two regiments of paratroops to Italy for the operation and the manufacture of some new equipment.[164] German airborne forces began to train, plan and prepare for the operation, tentatively scheduled for July.[165] Drop zones were selected on Malta, and Student trained the paratroops and aircrews relentlessly to improve dropping accuracy and speed of drop zone assembly.[166] However, Hitler did not trust the additional Italian forces required for the mission, particularly the Italian Navy, and was unwilling to suffer the casualties of "another Crete."[167] He cancelled the operation and the paratroops were sent instead to North Africa to reinforce Rommel. Malta remained in British hands, and RAF aircraft operating from Maltese airfields continued to choke off the *Afrika Korps*' supply lines across the Mediterranean until the surrender of the *Afrika Korps* in May 1943.[168] In the first six months after the Crete operation alone, 280,000 tons of military cargo were sunk en route to Rommel's forces.[169]

---

[162] Weeks, John. *Airborne Equipment*. Hippocrene Books, NY. 1976. pp. 76-81.
[163] Dugelby, Thomas and Stevens, Blake. *Death From Above, the German FG42 Paratroop Rifle*. Collector Grade Publications. Toronto. 1990. p. 80. The U.S. M-60 machine gun copied the principle. Production difficulties caused only 7,000 *FG42*'s to be produced.
[164] Schlicht, Adolf and Angolia, John. *Uniforms and Traditions of the Luftwaffe, Volume 2*. R. James Bender Publishing. San Jose. 1997. p. 402.
[165] Farrar-Hockley, p. 117.
[166] Ibid., pp. 114-115.
[167] Student & Götzel, pp. 350-362. The Italian airborne unit assigned to the Malta operation, the *Folgore Division*, had a good combat record but did not make a combat jump during the war. The *Folgore* was one of the few Italian units to remain fighting with the *Wehrmacht* after the collapse of the Mussolini regime and during the RSI period.
[168] Magenheimer, p. 282.
[169] Cooper, Matthew. *The German Army 1933-1945*. Stein and Day, New York. 1978. p. 360.

Student remained optimistic, however, about the future of the paratroops. In a classified report directly to *Reichmarschall* Göring in November 1942, a year and a half after the last airborne operation, he elaborated the correction of a number of earlier problems, writing: "Great progress has been made since Crete." In his report, Student discussed the new doctrine of jumping with weapons, improvements in dropping artillery, improved parachutes with faster opening harnesses, the incorporation of night jump training at the parachute schools, and experiments in jumping from other (faster) aircraft as well as from gliders. "Larger gliders now enable them to carry the heaviest weapons into battle with them," he noted. Yet he was painfully aware that his brainchild remained in a tactical backwater, concluding his report:

> The High Command has not undertaken any airborne operations since Crete. Now, when the stiffening of enemy resistance on all fronts reveals the possibility of long drawn out and bloody fighting, and when the striking power of our paratroops is immense, is the time to surprise the enemy by airborne operations at key points in his rear. Only thus can a contribution be made towards rapid and decisive victory.[170]

But for more than two years, between May 1941 and July 1943, while the *Luftwaffe* still remained a force to be reckoned with, no German parachute operations were conducted, even on the Russian Front, where airborne assaults could have been highly effective. Nevertheless, Student remained with his command and continued to advocate for them. "I pressed the idea on [Hitler] repeatedly," he said after the war, "but without avail."[171]

Part XI: A Fighting Reputation

Despite the cancellation of *Operation Herkules*, the airborne branch continued to grow in size as an all-volunteer force until the end of the war, although proportionally few of the men joining the new

---

[170] Student, General Kurt. *The Future of German Paratroop and Airborne Operations*. November 10, 1942 report to *Reichsmarshall* Göring. U.S. Air Historical Branch Translation No. VII/18 of January 14, 1947, declassified September 22, 1972.

units after 1942 were actually trained parachutists. Although he initially dismissed Student's reports that the Allies were forming airborne units in earnest as "latrine rumors",[172] Hitler saw the proof of it for himself in the Allied landings in North Africa. Too late, he rethought the matter. On February 13, 1943, he authorized the formation of a 2nd Parachute Division around a cadre of experienced paratroops from the 1st Parachute Division[173] to be commanded by the redoubtable General Ramcke, the man who reinforced the *Sturm Regiment* at Maleme in 1941.[174] At the end of 1943, two additional divisions were created, and, in April 1944, another two were established. By the end of the war, a total of 11 parachute divisions had been authorized, although only the first six reached anything close to their TO strength. Eventually, two corps-level commands, the *1. and 2. Fallschirm-Korps,* and ultimately an Army command, the *1. Fallschirm-Armee*, were created for command and control.[175]

At the time of the Normandy invasion in the summer of 1944, approximately 160,000 men were serving in German airborne units[176] but, accounting for casualties, even with expanded training efforts, only about 15 percent of them were actually trained parachutists. The trained paratroops were scattered throughout the various divisions and independent commands, and this policy led directly to one of the greatest impediments to conducting an airborne operation once Hitler's ban was lifted. Even if a battalion-sized operation was contemplated, for example, on average only about 15% of the men in the battalion were actually trained as paratroops.[177] This contrasts sharply with Allied practice. Nothing in the historical documentation of the period or in Student's memoirs suggests that he made an effort to consolidate the trained paratroops into a single regiment after

---

[171] Hart, p. 139.
[172] Farrar-Hockley, p. 103.
[173] Busch, p. 74.

[174] Thomas and Wegmann, p. 420.
[175] Ibid., pp. 411-442.
[176] Farrar-Hockley, p. 134.
[177] The percentages varied from unit to unit, and were significantly higher in the 1st and 2nd Parachute Divisions. Allied intelligence estimated after the war that 50% of the men in the 1st *Fallschirm Division* and 30% of the men of the 2nd *Fallschirm*

Crete.[178] It may be that the military manpower crisis within the Third Reich made even this impossible, but despite their defensive prowess, the squandering of men with such expensive training in the role of conventional infantry was nevertheless a curious and wasteful policy.

The German paratroops did fight tenaciously as infantry throughout the final four years of the war, notably in defense of Brest, where Ramcke's men inflicted 10,000 casualties on General Troy Middleton's VIII Corps,[179] and at Monte Cassino, which cost the Allies 100,000 casualties to take.[180] Reflecting on the battles of Monte Cassino after the war, British Field Marshall Lord Alexander called the *Fallschirmjäger* "the toughest, hardest fighting troops the world has ever seen."[181] General Sir John Hackett, himself a paratrooper, described their performance as "outstanding" and wrote that he could "personally testify wholeheartedly to their high quality as soldiers and as men."[182] The view of senior Allied commanders like Alexander, Hackett and Middleton was generally mirrored by British and American foot soldiers at the front. After Sicily, the British called them the "Green Devils."[183] They were regarded as tough opponents by G.I.'s as well, as historian Peter Schrijvers discovered in U.S. archives:

> After the African campaign, the paratroops became the most admired German soldiers. An officer of the 29th Infantry Division...claimed they 'were among the most dangerous fighting men of the war.' (The G.I.'s) described the paratroopers as 'fierce, stubborn, crafty, ingenious warriors' who were 'almost sullen in their determination.'[184]

Division were parachute trained. By the time of the Sicily drop, however, German sources indicate 100% of 1st *Fallschirm Division*'s combat elements were still parachute qualified. *Department of the Army Pamphlet 20-232,* p. 13.

[178] Student did suggest that transport squadrons with training and experience in airborne operation should be retained in reserve, but in the increasingly desperate situation in Germany after 1942, this was impossible. The squadrons were needed everywhere. As a stopgap measure, Student created a small group of "Paratroop Observers" (*Fallschirmjäger-Beobachter*) with combat jump experience to ride in the transports to assist the flight crews in reaching the drop zone and dispatching the paratroops, a classic example of his remarkable "outside the box" ability to improvise.

[179] D'Este, Carlo. *Decision in Normandy.* E. P. Dutton, NY. 1983. p. 434n. After the war Middleton commented that "during the whole of my combat career in two wars, I never met better fighting troops than the [German paratroops] at Brest."
[180] Ellis, John. *Cassino, the Hollow Victory.* McGraw Hill, NY. 1984. p. 469.
[181] Mitcham, Samuel and von Stauffenberg, Friedrich. *The Battle of Sicily.* Orion Books. New York. 1991. p. 141.
[182] Barnett, p. 474.
[183] Nasse, Jean-Yves. *Green Devils!* Histoire & Collections. Paris. 1997.

Unlike the *Waffen-SS*, Schrijvers found, "Most G.I.'s…looked upon the paratroopers as dedicated professionals rather than fanatics and grudgingly admitted a kind of respect."[185]

Their appreciation of the fighting ability of the *Fallschirmjäger* was in fact shared by Hitler himself. In a conversation with Albert Speer, he once called the paratroopers "the toughest fighters in the German army, tougher even than the *Waffen SS*."[186] Their reputation made them victims of their own success. They were rushed from one sector to another as each new defensive crisis appeared, and were referred to only half-jokingly as "Hitler's firemen."[187]

Part XII: Final Airborne Operations 1943-1945

Germany did mount seven airborne operations after Crete. Only one of them, however, was in greater strength than a reinforced battalion, and none of them had any real significance to the course of the war. These can be briefly summarized as follows (see also Appendix C, Chart of Post-Crete Airborne Operations):

Sicily, July 1943. Parachute forces in about regimental strength (elements of FJR 3 and FJR4) jumped behind German lines in reinforcement of the paratroop units already on the island. This use of paratroops followed the model of the Narvik operation in 1940. The majority of the reinforcements dropped safely and moved up to the front. It was a fast and efficient means of

---

[184] Schrijvers, Peter. *The Crash of Ruin, American Combat Soldiers in Europe during World War II*. New York University Press. New York. 1998. p. 64.
[185] Ibid.
[186] Clark, Alan. *The Fall of Crete*. Anthony Bond, Ltd. 1962. p. 49.
[187] Kuhn, pp. 205, 212.

reinforcing troops in difficult terrain, but it was not a parachute assault in the sense of seizing a defended objective.[188]

Elba, September 1943. In early September 1943, the 7,000 man Italian garrison on the island of Elba[189] indicated its willingness to surrender to British forces. To retain control of the island, a battalion of *Luftwaffe* paratroops (III./FJR7) was scrambled and dropped on the southwest corner of the island on September 17. Although resistance was anticipated, the Italian garrison surrendered without firing a shot. Elba remained in German hands until late in the war. The operation was professionally executed, but the island was strategically insignificant.[190]

Monte Rotondo, September 1943. One of the most rapidly planned and executed airborne operations of the war, the jump on Monte Rotondo by *II. Bataillon* of *FJR6* was also one of the most successful carried out by the *Fallschirmjäger*. When the Italian government collapsed in September, 1943, the Germans moved very quickly to retain territorial control. Paratroops moved into Rome on the ground, and under the command of Major Gericke, the acting commander of the *Sturm Regiment* on Crete, II/FJR6 jumped on the Italian Army headquarters complex at Monte Rotondo north of Rome on September 9. After a sharp firefight, Gericke and his men seized the Italian general staff and negotiated a cease-fire by Italian forces against the Germans throughout the country. The operation stabilized the situation in Italy and enabled the Germans to quickly consolidate control countrywide without further resistance. It was a very effective use of paratroops.

---

[188] Mitcham and von Stauffenberg, pp. 160-5.
[189] Elba is just off the Italian west coast, halfway between Rome and Genoa. It was the place of Napoleon's first exile from 1814-15.

Gran Sasso, September 1943. The *Luftwaffe* parachute test section, the *Fallschirm-Lehr Battalion*, mounted a glider assault on a mountaintop hideaway in Italy, normally accessible only by cable car, to kidnap Mussolini from pro-Allied Italian forces. It was one of the most successful commando operations of World War II.[191] An *SS* officer named Otto Skorzeny accompanied *Luftwaffe Oberleutnant* von Berlepsch on the mission as an observer, jumped into the small plane carrying Mussolini at the last minute, then claimed all the credit for the operation.[192] In reality, Skorzeny had nothing to do with the planning and execution of the mission, but the self-promoted myth of "Skorzeny's great success" has continued to the present day.[193] The operation had significant propaganda value and made possible a brief Mussolini rump regime, the "Republic of Salo," or RSI, in northern Italy.[194]

Leros, November 1943. This was a combined airborne and amphibious operation conducted to recapture one of the Dodecanese Islands from the British. A *Luftwaffe* parachute battalion (*II./FJR2*) and the army special forces parachute battalion of the Brandenburg Regiment[195] dropped in daylight and secured the waist of the hourglass-shaped island. The island was retaken after a week of ferocious fighting with the British garrison.[196] It had no effect on the war.

---

[190] Stimpel, Hans Martin. *Die Deutsche Fallschirmtruppe 1942-1945, Einsätze auf Kriegsschauplätzen im Süden*. Verlag E.S. Mittler & Sohn GmbH. Hamburg. 1998. p. 254-5.
[191] McNab, Chris, ed. *German Paratroopers, The Illustrated History of the Fallschirmjäger in World War II*. MBI Publishing. Osceola, WI. 2000. p. 145.
[192] Stimpel, Hans Martin. *Die Deutsche Fallschirmtruppe 1942-1945, Einsätze auf Kriegsschauplätzen im Süden*. Verlag E.S. Mittler & Sohn GmbH. Hamburg. 1998. pp. 150-4. See also Lucas, *Storming Eagles*, p. 107.

[193] When he heard Skorzeny's claim on a German radio broadcast, The *Luftwaffe* officer commanding the operation, Major Otto Harald Mors, was so angered by this usurpation of the laurels due his own men that he took the unusual step of invoking his right to complain directly to the *Luftwaffe* High Command (OKL). He was told the propaganda statement was approved by Hitler himself. Farrar-Hockley, p. 126.
[194] Murray and Millett, p. 383.
[195] Kurowski, Franz. *The Brandenburgers Global Mission*. J.J. Federowicz Publishing, Winnipeg. 1997. pp. 256-262.
[196] Lucas, pp. 108-110.

Drvar, May 1944. In the fall of 1943, the *Waffen-SS* raised a battalion of paratroops, partly from penal companies, designated as *SS-Fallschirmjäger Bataillon 500*. On May 25 1944, about 850 paratroops in two waves jumped on Marshall Tito's headquarters near Drvar in an attempt to capture or kill the Yugoslav partisan leader. The *SS* paratroops missed Tito by minutes,[197] but caused a significant disruption in partisan activities in the region for several months. Had they succeeded, the post-war history of the Balkans would have been rather different. The *SS* parachute battalion finished the war fighting as infantry on the Russian Front.[198]

Ardennes, December 1944. Conceived as part of Germany's winter offensive in 1944 in the Ardennes,[199] the final German airborne operation of WWII was a disaster from its inception. A 1,000-man paratroop battle group commanded by General von der Heydte was tasked with the seizure of a number of vital road features ahead of the advancing *Panzer* columns, but was scattered instead over hundreds of miles.[200] One company was even dropped over Bonn.[201] The one-week preparation time was inadequate, as the few remaining parachute trained men were scattered in units all over Europe, and there were essentially no remaining Ju52 pilots or crews with experience in parachute drops. The operation was a complete failure, except that their presence tied U.S. rear area security into knots, and a rumor that German paratroops were planning to kidnap General Eisenhower caused two U.S. infantry divisions to be diverted to guard his HQ a hundred miles behind the front, where SACEUR was a virtual prisoner in his own offices for nearly a month, a reflection of the Allied respect for the *Fallschirmjäger*, even at that late date.

---

[197] Student and his staff were frustrated that they had not been consulted in the planning for the first (and only) *SS* parachute operation, as their experience might have prevented several costly errors. The phrase *"Ihn knapp verfehlt"* ("just missed him") became a kind of sarcastic joke among Student's staff afterwards. Farrar-Hockley, p. 134.
[198] Milius, p. 17.
[199] Department of the Army *Historical Pamphlet 20-232*. pp. 21-22.
[200] Lucas, p. 149.
[201] Whiting, Charles. *Ardennes, the Secret War*. Stein and Day. New York. 1985. p. 105.

Discounting the *SS* operation, which *Luftwaffe* planners were unaware of, five out of the six *Luftwaffe* parachute operations mounted after Crete were fully successful. Only the hopeless final drop in the Ardennes was a tactical failure, but given the circumstances it is a tribute to the capability and professionalism of Student's staff that they managed to mount one at all. This certainly reflects a high degree of competence in both planning and execution in the airborne branch in 1943 and 1944, despite the obviously declining fortunes of both Germany and the *Luftwaffe*. The five successful operations also demonstrated that Student had absorbed the tactical and doctrinal lessons of Crete and corrected all of them by 1943. Missions and objectives were tailored to the paratroop's capabilities and limitations, and paratroopers now always jumped with their weapons, with increased firepower, and on safer drop zones. On Sicily, they intentionally dropped within their own lines. On Gran Sasso and Elba, no combat was required, and at Monte Rotondo, only a sharp but brief exchange of fire was necessary to establish control. Indeed, Monte Rotondo, in particular, was a model example of a successful airborne raid to achieve a strategic objective.

Nevertheless, the ability of the German airborne to significantly influence events on the battlefield had come and gone. By mid-1943, Germany was in retreat on all fronts, and no longer had the transport aircraft or air power to mount a genuinely potent airborne operation, nor the offensive striking power to exploit one. What began as a promising revolution in military tactics in 1936 ended under the steamroller of Allied military power.

Part XIII: What Might Have Been

Practically an entire cottage industry has grown up around the "what-ifs" of World War II. There have been numerous books on the subject in addition to movies and even computer games. Despite this, it still might be useful to ask whether the airborne could have given Hitler the key to victory in World War II before Pearl Harbor and American materiel superiority made the war's outcome inevitable?

If Hitler had put aside his obsession with the Soviet Union[202] and pursued his war with Great Britain to a conclusion first, and *if* Hitler had listened to his admirals and followed a Mediterranean strategy[203] after abandoning plans for *Operation Seelöwe*, the answer is "yes." A German airborne assault on Malta instead of Crete would have secured that strategic key to the Mediterranean and secured Rommel's lifeline to North Africa. As Crete and later Elba both demonstrated, the use of paratroops enabled German land-based forces to neutralize British naval supremacy as a factor in amphibious operations. If after taking Malta, Hitler had then vigorously pursued the North African campaign instead of launching *Operation Barbarossa*, he could very likely have driven the British from the Middle East and secured the oilfields of Iran. As historian Bevin Alexander noted:

> With this decision (to take Crete rather than Malta), Adolf Hitler lost the war. The assault on Crete guaranteed two catastrophes for Germany; it limited the Mediterranean to peripheral public relations goals, and it turned German strength against the Soviet Union while Britain remained defiant, with the United States in the wings.[204]

Indeed, Hitler bumbled the North African campaign from beginning to end, not just in his failure to seize Malta, but moreover in his failure to prioritize resupply to Rommel's forces and above

---

[202] Hart, p. 137.
[203] Magenheimer, p. 69.

all in his lack of strategic vision in the theatre. The resource allocations to the two fronts tell the story: One hundred and forty-eight divisions were committed to *Barbarossa*; only three were initially committed to North Africa.[205] When the British launched their "Crusader Offensive" in November 1942, Rommel still had only 249 German made tanks, of which 75 were obsolete *PzKw I's* and *II's*, against 1238 British machines.[206]

Victory in World War II was well within Hitler's grasp before 1942, and only a series of serious blunders on his part, chief among them his prosecution of the war against the Soviet Union, ultimately enabled the forces of democracy to prevail. The point here is not idle speculation, but rather the fact that the airborne weapon, if used properly, could have been a force multiplier and an enabler of decisive victory at the strategic level. How much influence Student's advocacy of Crete over Malta actually had on Hitler's decision and the outcome of the war is debatable, given Hitler's obsession with the Soviet Union.

Part XIV: Final Thoughts
"Great things have been effected by a few men well conducted."[207]

The Germans attempted a total of 15 operational parachute drops during World War II. A remarkable 80 percent of them (12 operations) were fully successful. 7 percent of them (1 operation) were aborted due to inclement weather, and 13 percent (Drvar and the Ardennes) were unsuccessful. 86 percent of the *Luftwaffe* operations were fully successful, and their sole operational failure, the Ardennes, was conducted at a time when Germany was at the end of its

---

[204] Alexander, Bevin. *How Hitler Could Have Won World War II*. Crown Publishers. New York. 2000. p. 63.
[205] Magenheimer, p. 75.
[206] Cooper, *The German Army 1933-1945*, p. 374.

[207] George Rogers Clarke, 1777. (Albert, Robert. *George Rogers Clark and Winning the Old Northwest.* National Park Service Pamphlet, Government Printing Office, Washington, DC. p. 3)

tether. This is a considerably better record than the Allies, with all their overwhelming materiel superiority, were able to accomplish.

The statistics also show that the conventional historical wisdom on the German airborne branch, that Crete was the end of airborne operations, is an oversimplification. Crete was indeed the last large-scale parachute operation. However, despite a two-year hiatus from May 1941 to May 1943, Nazi Germany conducted the same number of airborne operations *after* Crete (seven) as it had before Crete (seven; for a summary of these operations, see Appendix B).[208] While the scale of these later operations was smaller, the success rate of the *Luftwaffe* airborne missions was substantially unchanged. 88 percent of the operations up to and including Crete accomplished their assigned missions while 83 percent of those conducted after Crete did so. It would be more correct to say that Germany used airborne operations with great success early in the war, undertook a two-year re-evaluation and adjustment program after Crete, then used them successfully again at the end of the war on the smaller scale dictated by Germany's defensive posture on all fronts and her rapidly shrinking resources.

The pioneering German effort in airborne operations 60 years ago still provides a wealth of insights applicable to parachute operations today, and in fact, the United States studied the matter closely after the war.[209] The first and most important of these observations is that air superiority is the *sine qua non* of parachute operations. Without mastery of the air, no paratroop operation can succeed in either the delivery or the exploitation phase. Furthermore, utilizing transport pilots unfamiliar with parachute operations in combat is a recipe for disaster.

---

[208] This includes the *SS* parachute operation at Drvar, which was conducted outside the control of the *Wehrmacht*.

The second observation is that there will always be missions in war and military operations other than war that a highly-trained group of tough, elite light infantry capable of arriving from the air at any point, can best undertake via vertical envelopment. Such units, whether they arrive by parachute or helicopter, have utility in seizing tactical objectives as part of a coordinated, combined-arms offensive, but their limitations on the tactical level of operations must be respected. Both Montgomery and Student failed on one critical occasion to do this, and it was catastrophic for both.

The third observation is that paratroops are a "rich man's" weapon, as senior German officers emphasized after the war.[210] By this they meant that in wartime, only nations rich in men and materiel could afford to keep such highly trained specialist troops and trained air transport squadrons in the rear in readiness for potential operations and not in action at the front. When a nation of limited means and manpower is engaged in a major war, it can generally ill afford to keep its best forces and a portion of its air transport assets idle while awaiting the employment for which they are designed.

Finally, and most importantly, there will always be new weapons in war and new ways of waging war. In his restless urge to conquer and his instinct to defend, man has never ceased in his search for new methods of gaining superiority on the battlefield. Any nation which seeks to defend itself against outside enemies must explore fully every means of waging war which those enemies could use, even those which appear far-fetched or even preposterous, not merely those

---

[209] The U.S. Army's conclusions are discussed at length in Department of the Army Pamphlet No. 20-260, *The German Campaign in the Balkans*.

which fit its preconceptions of modern war. To embrace one new technique, such as building massive fortifications, and ignore another, such as vertical envelopment, is to invite disaster, as the French and Belgians learned to their cost in World War II.

Before the Second World War, many senior officers in the countries which would become the Allies clung to obsolete conceptions of war. It was far easier to hold on to the old ways than to adapt to new possibilities. Yet many of the cherished elements of war fighting, such as battleships and horse cavalry, were no longer viable on the battlefield. Today, airborne warfare, the way of war that was ushered into our consciousness by General Kurt Student, remains in use in dozens of countries around the world, including the United States, which maintains the 82nd Airborne Division in readiness for worldwide contingencies. Since Student's day, helicopters have replaced gliders, but air assault forces and paratroops remain very much a part of modern military thinking in the developed world. Student's ultimate insight for all defense planners, however, lies not in his revolutionary tactics of vertical envelopment. Rather it is in his vision, and what it reminds us of: The never-ending need to search the minds of our enemies for the next means of bypassing our defenses.

---

[210] Department of the Army. *Historical Study No. 20-232*, p. 43.

German Parachute Operations Through Crete

88% Success Rate (12% Weather Aborted)

(Opp= opposed landing anticipated, Tacon = tactical control, LW = *Luftwaffe*)

Airborne Operation	Code Name	Date	Purpose	Type	Opp	Troops	Tacon	Success	Comments
Falster-Fyn Bridge	*Weserübung*	April 1940	Seize Bridge	Parachute	Yes	160	LW	Yes	No casualties
Oslo-Forebu	*Weserübung*	April 1940	Reinforcement	Parachute	No	320	LW	No	Aborted, weather
Gudbrandsal	*Weserübung*	April 1940	Hold Crossroads	Parachute	Yes	160	LW	Yes	50% casualties
Narvik	*Weserübung*	May 1940	Reinforcement	Parachute	No	320	LW	Yes	Some army personnel jumped without parachute training
Eben Emäl	*Fall Gelb*	May 1940	Seize Fortress	Glider	Yes	80	LW	Yes	800 POWs taken
Holland, Various	*Fall Gelb*	May 1940	Seize Objectives	Parachute	Yes	1000	LW	Yes	Moerdijk Bridge, Dordrecht Bridge, Waalhoven airfield seized
Corinth	[none]	May 1941	Blocking force	Parachute	Yes	1500	LW	Yes	2,500 British POWs taken for 63 German KIA
Crete	*Merkur*	May 1941	Seize Island	Parachute & Glider	Yes	8300	LW	Yes	50% casualties

German Post-Crete Parachute Operations

83% *Luftwaffe* Success Rate (Less Failed SS Kidnapping Special Operation at Drvar)
(Opp= opposed landing anticipated, Tacon = tactical control, LW = *Luftwaffe*)

Airborne Operation	Code Name	Date	Purpose	Type	Opp	Troops	Tacon	Success	Comments
Sicily	[none]	July 1943	Reinforcement	Parachute	No	2000	LW	Yes	First airborne vs. airborne battle in history
Monte Rotondo	*Schwartz*	September 1943	HQ Seizure	Parachute	Yes	600	LW	Yes	Captured Italian General HQ
Elba	[none]	September 1943	Seize island	Parachute	Yes	600	LW	Yes	Captured 7,000-man Italian garrison
Gran Sasso	*Eiche*	September 1943	Special Op	Glider	Yes	80	LW	Yes	Kidnap Mussolini
Leros	*Leopard*	November 1943	Blocking Force	Parachute	Yes	1000	LW	Yes	Combined *Luftwaffe* and Army Operation
Drvar	*Rösselsprung*	May 1944	Special Op	Parachute & Glider	Yes	850	SS	No	SS operation to kidnap or kill Tito
Ardennes	*Stösser*	December 1944	Seize Objectives	Parachute	Yes	1000	LW	No	800 of 1,000 parachutists captured

Bibliography

<u>Books</u>

Ailsby, Christopher. *Hitler's Sky Warriors*. Brassey's, Inc. Dulles, VA. 2000.

Albert, Robert. *George Rogers Clark and Winning the Old Northwest*. National Park Service Pamphlet, Government Printing Office, Washington, DC. Undated.

Alexander, Bevin. *How Hitler Could Have Won World War II*. Crown Publishers. New York. 2000.

Alman, Karl. *Sprung in die Hölle*. Erich Pabel Verlag GmbH & Co. Rastatt. 1964.

Arnold, General H.H. *Winged Warfare*. Harper & Brothers. New York. 1941.

Barnett, Correlli, ed. *Hitler's Generals*. William Morrow & Co. New York. 1989.

Bender, R. James and Law, Richard. *Afrikakorps*. R. James Bender Publishing. San Jose. 1973.

Bender, Roger. *Air Organizations of the Third Reich*. R. James Bender Publishing. San Jose. 1967.

Bennett, Ralph. *Ultra and Mediterranean Strategy*. William Morrow. New York. 1989.

Beevor, Anthony. *Crete, the Battle and the Resistance*. West View Press. Boulder, CO. 1994.

Brockdorff, Werner. *Geheim Kommandos des Zweiten Weltkrieges*. Weltbild Verlag GmbH. Augsburg. 1993.

Bullock, Alan. *Hitler and Stalin*. Vintage Books, Random House. New York. 1991.

Busch, Erich. *Die Fallschirmjäger Chronik 1935-1945*. Podzun Pallas Verlag. Friedberg. 1983.

Clark, Alan. *The Fall of Crete*. Anthony Bond, Ltd. 1962.

Cooper, Matthew. *The German Air Force 1933-1945, an Anatomy of Failure*. Jane's Publishing, Cooper and Lucas, Ltd. London. 1981.

Cooper, Matthew. *The German Army 1933-1945*. Stein and Day. New York. 1978.

Department of the Army. *Historical Study No. 20-232, Airborne Operations, A German Appraisal*. October 1951.

Department of the Army. *Pamphlet No. 20-260, The German Campaign in the Balkans (Spring 1941)*. November 1953.

D'Este, Carlo. *Decision in Normandy*. E. P. Dutton. New York. 1983.

Devine, Isaac. *Mitchell, Pioneer of Air Power*. Duell, Sloan and Pearce. New York. 1943.

Devlin, Gerard. *Paratrooper!* St. Martin's Press. New York. 1979.

Dugelby, Thomas and Stevens, Blake. *Death From Above, the German FG42 Paratroop Rifle*. Collector Grade Publications. Toronto. 1990.

Edwards, Roger. *German Airborne Troops*. Doubleday & Co. Garden City, New York. 1974.

Ellis, John. *Cassino, the Hollow Victory*. McGraw Hill. New York. 1984.

Farrar-Hockley, A.H. *Student*. Ballantine Books. New York. 1973.

Dr. Fink, Feri. *Der Komet auf Kreta*. Gelka-Druck und Verlags. Ettlingen. c. 1990.

Fraser, David. *Knight's Cross, A Life of Field Marshall Erwin Rommel*. Harper Collins Pub. UK. 1993.

Freedman, Maurice. *Unraveling Enigma*. Pen & Sword Books, Ltd. South Yorkshire, England. 2000.

Griess, Thomas, series ed. *The Second World War: Europe and the Mediterranean*. Department of History, United States Military Academy, West Point. Avery Publishing. New Jersey. 1989.

Hart, Liddel. *History of the Second World War*. Da Capo Press. New York. 1999.

Haupt, Werner. *Fallschirmjäger 1939-1945, Weg und Schicksal einer Truppe*. Podzun-Pallas Verlag. Friedberg. 1979.

Von der Heydte, Dr. Freiherr Friedrich August. *Daedalus Returned*. Hutchinson Publishing. London. 1958.

Hickey, Michael. *Out of the Sky*. Charles Scribner's Sons. New York. 1979.

Hill, Michael. *Black Sunday, Ploesti*. Schiffer Publishing. Atglen, PA. 1993.

Hoffman, LTC Jon. *Silk Chutes and Hard Fighting: U.S. Marine Corps Parachute Units in World War II*. History and Museums Division, Headquarters Marine Corps. Washington, DC. 1999.

Von Hove, Alkmar. *Achtung Fallschirmjäger*. Druffel-Verlag Freising. 1954.

Hoyt, Edwin. *Hitler's War*. Plenum Publishing. New York. 1988.

Kaloudis, Pantelis. *Crete May 1941*. Albion Scott Ltd. Brentford. 1981.

Keitel, Wilhelm. *The Memoirs of Field Marshall Wilhelm Keitel*, ed. Walter Gorlitz. Cooper Square Press. New York. 2000.

Kiriakopoulos, G.C. *Ten Days to Destiny*. Franklin Waltz. New York. 1985.

Knox, Macgregor. *Hitler's Italian Allies*. Cambridge University Press. Cambridge. 2000.

Kuhn, Volkmar. *German Paratroops in World War II*. Ian Allen Ltd. London. 1978.

Kurowski, Franz. *The Brandenburgers Global Mission*. J.J. Federowicz Publishing. Winnipeg. 1997.

Kurowski, Franz. *Der Kampf um Kreta*. Manfred Pawlak Verlagsgesellschaft. Herrsching. 1985.

Kurowski, Franz. *Deutsche Fallschirmjäger 1939-1945*. Edition Aktuell. Frankfurt. c. 1990.

Kurtz, Robert. *German Paratroops*. Schiffer Publishing. Atglen, PA. 2000.

Lucas, James. *Storming Eagles*. Arms and Armor Press. London. 1988.

McNab, Chris, ed. *German Paratroopers, The Illustrated History of the Fallschirmjäger in World War II*. MBI Publishing. Osceola, WI. 2000.

MacDonald, Callum. *The Battle of Crete*. Macmillan. New York. 1993.

Magenheimer, Heinz. *Hitler's War*. Cassell & Co. London. 1998.

Mason, Herbert Molloy, Jr. *The Rise of the Luftwaffe 1918-1940*. The Dial Press. New York. 1973.

Milius, Siegfried. *Fallschirmjäger der Waffen-SS im Bild*. Munin Verlag GmbH. Osnabrück. 1986.

Mitcham, Samuel W., Jr. *Hitler's Legions, The German Army Order of Battle*. Stein and Day. New York. 1985.

Mitcham, Samuel and von Stauffenberg, Friedrich. *The Battle of Sicily*. Orion Books. New York. 1991.

Mitchell, William. *Memoirs of World War I*. Random House. 1960.

Mrazek, James E. *The Fall of Eben Emael*. Self-published. Library of Congress no. 74-110765.

Murray, Williamson and Millett, Allan. *A War To Be Won*. The Belknap Press of Harvard University Press. Cambridge. 2000.

Nasse, Jean-Yves. *Green Devils!* Histoire & Collections. Paris. 1997.

Nowarra, Heinz. *German Gliders in World War II*. Schiffer Publishing. West Chester PA. 1991.

Otte, Alfred. *The HG Panzer Division*. Schiffer Publishing Ltd. West Chester PA. 1989.

Peters, Klaus. *Fallschirmjäger Regiment 3*. R. James Bender Publishing. San Jose. 1992.

Piehl, Hauptmann. *Ganze Männer*. Verlagshaus Bong & Co. Leipzig. 1943.

Price, Alfred. *Luftwaffe Handbook 1939-1945*. Charles Scribner's Sons. New York. 1977.

Quarrie, Bruce. *Airborne Assault*. Patrick Stephens. Somerset. 1991.

Von Roon, Arnold. *Die Bildchronik der Fallschirmtruppe 1935-1945*. Podzun Pallas Verlag. Friedberg. 1985.

Ryan, Cornelius. *A Bridge Too Far*. Touchstone Books. New York. 1974.

Schlicht, Adolf and Angolia, John. *Uniforms and Traditions of the German Army 1939-1945, Volume 1*. R. James Bender Publishing. San Jose. 1984.

Schlicht, Adolf and Angolia, John. *Uniforms and Traditions of the Luftwaffe, Volume 2*. R. James Bender Publishing. San Jose. 1997.

Schrijvers, Peter. *The Crash of Ruin, American Combat Soldiers in Europe during World War II*. New York University Press. New York. 1998.

Shulman, Milton. *Defeat in the West*. Secker and Warburg. London. 1947.

Stewart, I.McD.G. *The Struggle for Crete*. Oxford University Press. London. 1966.

Stimpel, Hans Martin. *Die Deutsche Fallschirmtruppe 1942-1945, Einsätze auf Kriegsschauplätzen im Süden*. Verlag E.S. Mittler & Sohn GmbH. Hamburg. 1998.

Student, General Kurt, and Götzel, Hermann. *Generaloberst Kurt Student und Seine Fallschirmjäger*. Podzun Pallas Verlag. Friedberg. 1980.

Thomas, Franz and Wegmann, Günther. *Die Ritterkreuzträger der Deutschen Wehrmacht 1939-1945, Teil II: Fallschirmjäger*. Biblioverlag. Osnabrück. 1986.

Time Life Books ed. *Barbarossa*. Time Life Books, Inc. Alexandria VA. 1990.

Time-Life Books ed. *The Conquest of the Balkans*. Time-Life Books, Inc. Alexandria VA. 1990.

Veranow, Michael, ed. *The Third Reich at War*. Galahad Books. New York. 1999.

Weeks, John. *Airborne Equipment*. Hippocrene Books. New York. 1976.

Whiting, Charles. *Ardennes, the Secret War*. Stein and Day. New York. 1985.

Winterstein, Ernst and Jakobs, Hans. *General Meindl und Seine Fallschirmjäger*. Verlagsbuchbinderei Ladstetter GmbH. Hamburg. c. 1990.

Magazine Articles

Infantry Journal. "Air Infantry Training." July-August issue, 1940.

Reports

Student, General Kurt. *The Future of German Paratroop and Airborne Operations*. November 10, 1942 report to Reichsmarshall Göring. U.S. Air Historical Branch Translation No. VII/18 of January 14, 1947, declassified September 22, 1972.

German Army Research Report, *Der Luftlandeangriff auf Kreta*. Dokumentationszentrum der Bundeswehr, Bonn, Film Nr. 984.

Headquarters, United States Army Air Forces in Europe (Office of the Assistant Chief of Staff, A-2) Report of October 15, 1945, *Air Staff Post Hostilities Intelligence Requirements on German Air Force: Tactical Employment Troop Carrier Operations*. Declassified January 31, 1958.

British Intelligence M.I.14 Report "*M.I.14(j)/su/28/44 Training and Tactics of Parachutists, Winter 1943/44*," National Archives microfiche 512.6312-29, declassified.

Interviews

Sergeant Ernst Simon, *Fallschirmjäger Regiment 1*. Interview by the author, tape recording, Hamburg, Germany, 1995.

Lieutenant Colonel Wilhelm Reinhardt, *Fallschirmjäger Regiment 1*. Interview by the author, tape recording, Bruchsal, Germany, 1996.

Sergeant Major Martin Junge, *Luftlande Sturm Regiment*. Interview by the author, tape recording, Frankfurt, Germany, 1993.

Other Resources

Author's terrain walk, Crete, June 1986.

www.ingramcontent.com/pod-product-compliance
Lightning Source LLC
Chambersburg PA
CBHW081300170426
43198CB00017B/2855